ROMAN MOSAICS

IN THE J. PAUL GETTY MUSEUM

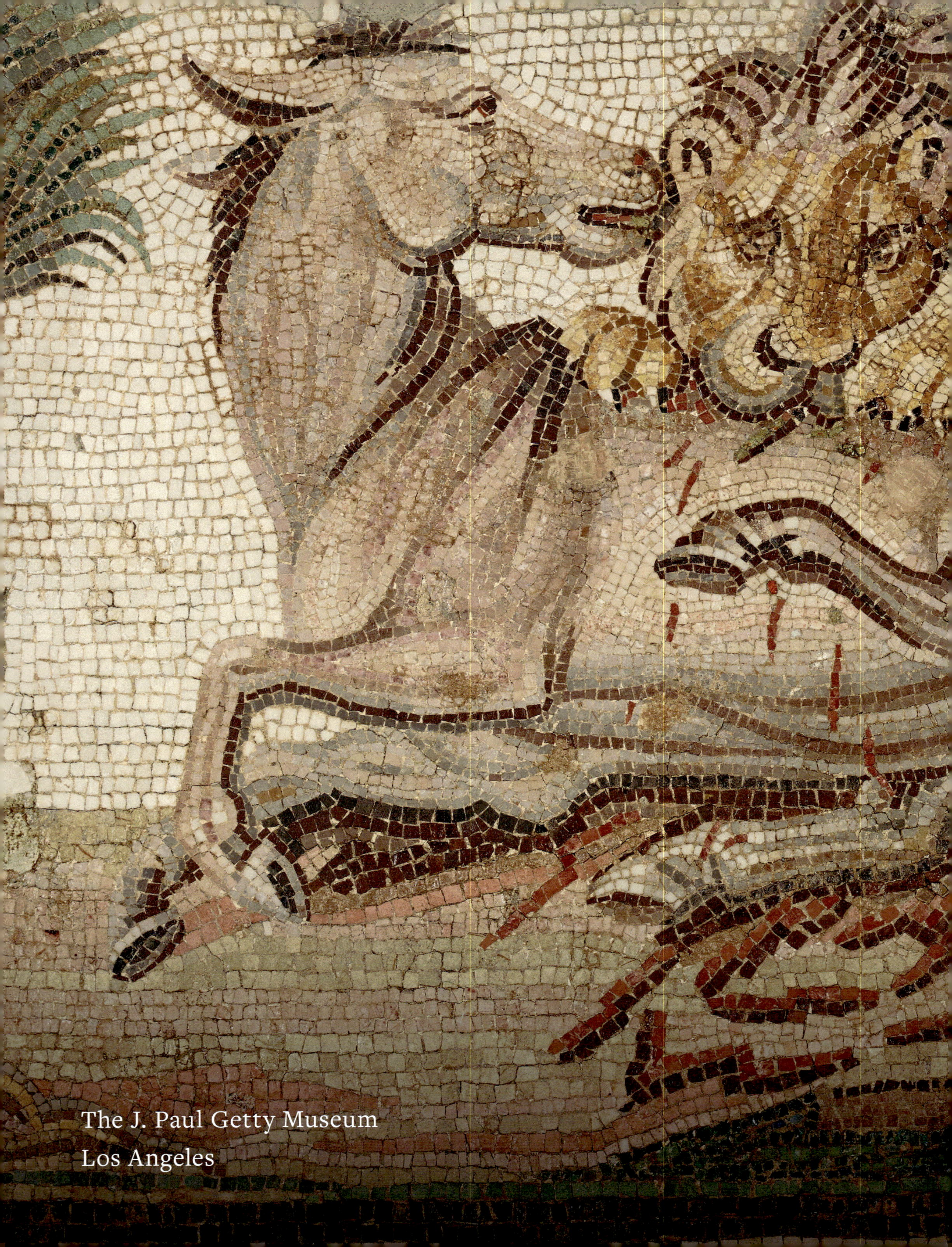

The J. Paul Getty Museum
Los Angeles

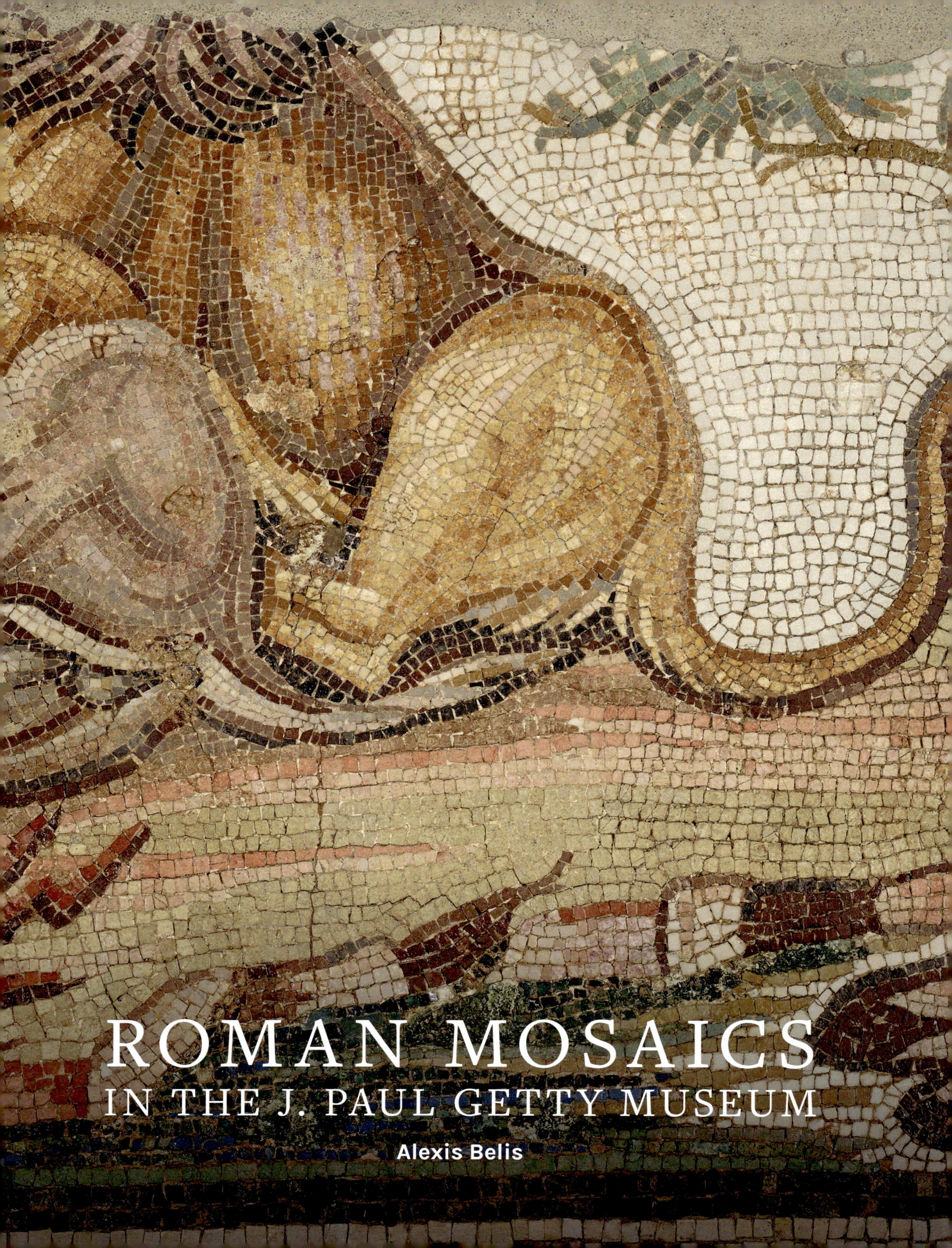

ROMAN MOSAICS
IN THE J. PAUL GETTY MUSEUM

Alexis Belis

The free online edition of this catalogue, available at http://www.getty.edu/publications/romanmosaics, includes zoomable high-resolution photography, embedded glossary terms and additional comparative images, and interactive maps drawn from the Digital Atlas of the Roman Empire. Also available are free PDF, EPUB, and MOBI downloads of the book, CSV and JSON downloads of the object data from the catalogue, and JPG and PPT downloads of the main catalogue images.

First edition, 2016
Last updated, September 19, 2017
https://www.github.com/gettypubs/romanmosaics

Published by the J. Paul Getty Museum, Los Angeles
Getty Publications
1200 Getty Center Drive, Suite 500
Los Angeles, California 90049-1682
www.getty.edu/publications

Nola Butler, *Project Editor*
Rachel Barth and Nola Butler, *Copy Editors*
Elizabeth Chapin Kahn, *Production*
Stephanie Grimes, *Digital Researcher*
Eric Gardner, *Designer and Developer*
Greg Albers, *Project Manager*

Distributed in the United States and Canada by the University of Chicago Press
Distributed outside the United States and Canada by Yale University Press, London

Library of Congress Cataloging-in-Publication Data
Names: Belis, Alexis, author. | J. Paul Getty Museum, issuing body.
Title: Roman mosaics in the J. Paul Getty Museum / Alexis Belis ; with contributions by Nicole Budrovich, Christine Kondoleon, Kenneth Lapatin, and Sean Leatherbury.
Description: Los Angeles : J. Paul Getty Museum, [2016] | ©2016 | Includes bibliographical references and index.
Identifiers: LCCN 2016003875 (print) | LCCN 2016011119 (ebook) | ISBN 9781606064979 (pbk.) | ISBN 9781606064986 (epub) | ISBN 9781606064993 (online)
Subjects: LCSH: Mosaics, Roman—Catalogs. | Pavements, Mosaic—Catalogs. | J. Paul Getty Museum—Catalogs. | Mosaics—California—Los Angeles—Catalogs. | Pavements, Mosaic—California—Los Angeles—Catalogs.
Classification: LCC NA3770 .J2 2016 (print) | LCC NA3770 (ebook) | DDC 738.50937/07479494—dc23
LC record available at http://lccn.loc.gov/2016003875

Front cover: Detail of Mosaic Floor with Head of Medusa (cat. 1)
Back cover: Detail of Mosaic Panel with Peacock Facing Left (cat. 9)
pp. i–ii: Detail of Mosaic of a Lion Attacking an Onager (cat. 5)
p. viii: Detail of Mosaic Floor with Combat between Dares and Entellus (cat. 4)
p. 66: Detail of Mosaic Floor with Orpheus and Animals (cat. 3)

Contents

Director's Foreword

Mosaics are one of the most distinctive and widespread forms of Roman art to survive from antiquity. The majority of them covered floors and, consequently, durability was a primary reason for creating these expensive, labor-intensive compositions in colored stones and glass, imitating what could have been achieved much more easily in paint. Apart from being merely functional, floor mosaics were a significant artistic medium for the Romans. *Roman Mosaics in the J. Paul Getty Museum* examines several examples in the Getty's collection with detailed narrative scenes and elaborate decorative patterns that adorned reception areas and dining rooms in private villas—spaces in which the owner could display his wealth, cultural sophistication, and artistic preferences. Beyond domestic settings, other significant mosaics in the exhibition embellished the interiors of a variety of public buildings, including baths, temples, and churches.

This catalogue includes all the examples of Roman mosaic art in the Getty's collection. From his earliest days of collecting antiquities in Rome in 1939, J. Paul Getty expressed an interest in this medium, commenting in his diaries on various pieces he saw in and around Rome, including those at the Musei Capitolini and the Museo Nazionale Romano—Terme di Diocleziano. He also visited the Vatican mosaic factory in Rome and, impressed by its beautiful mosaics, purchased one depicting a bowl of fruit. In his 1965 memoir, *The Joys of Collecting*, Getty recalls with pride how he came into possession of his first ancient mosaic, the Gallo-Roman Mosaic Floor with Orpheus and Animals (see cat. 3), in 1949. Getty was fascinated by Greek and Roman culture and, in the manner of the Romans, contracted a skilled artisan to install the mosaic over the existing floor in the "Roman Room" of his new museum—the Spanish Colonial ranch house in which his collection was displayed until the Getty Villa opened in 1974. Getty describes how he set one of the stones in the mosaic floor himself. Although many of the mosaics that have entered the collection over the years have been on view at the Getty Villa, some examples in this catalogue have never before been exhibited to the public or published.

Arranged geographically by regions of the Roman Empire, the catalogue entries situate each mosaic within a broad stylistic and typological framework and illuminate the context of its discovery. A number of the mosaics in the Getty's collection were included in the exhibition *Roman Mosaics across the Empire*, held at the Getty Villa in the spring of 2016. This publication was produced online and via print-on-demand to accompany the exhibition and make immediately available recent scholarly research on the collection. It is one of a new series of web-based catalogues of the collection of Greek, Roman, and Etruscan art at the Getty Villa. The author, contributors, and staff of Getty Publications are to be commended for realizing this innovative and accessible guide, which we hope will inspire more of our visitors to study and enjoy this important aspect of Roman art.

Timothy Potts
Director
The J. Paul Getty Museum

Acknowledgments

I extend my sincere thanks to the staff of the many departments in the J. Paul Getty Museum who were involved in the production of this catalogue, and in particular to Jeffrey Spier, senior curator of antiquities, for his support in completing this project. I am especially grateful to the other authors—Christine Kondoleon, Kenneth Lapatin, Nicole Budrovich, and Sean Leatherbury—who contributed their scholarly knowledge of the mosaics in the Getty's collection, and to Judith Barr for her ongoing research on the provenance of the mosaics. I would also like to acknowledge the contributions of André Girod and the association of Villa Laurus en Luberon, who provided invaluable material on the archaeological site near Villelaure, where the Getty Museum's mosaic of Dares and Entellus was discovered.

Special thanks are owed to the staff of the Villa Imaging Studio, especially Rebecca Truszkowski, who photographed nearly all of the mosaics for this publication, and imaging technician Benjamin Goddard, with the support of senior photographer Tahnee Cracchiola and assistant imaging technician Rosanna Chan. I extend heartfelt gratitude to Eduardo Sanchez, Jeffrey Maish, Erik Risser, Marie Svoboda, and Susan Lansing Maish in the Antiquities Conservation department, as well as conservation interns Sara Levin and Ellie Ohara, for their exceptional work in preparing the mosaics for not only photography but also display in the exhibition *Roman Mosaics across the Empire*. I am grateful also to Nola Butler, Greg Albers, Eric Gardner, Elizabeth Kahn, Rachel Barth, and all of the staff of Getty Publications who made this catalogue a reality.

Alexis Belis
Assistant Curator, Antiquities
The J. Paul Getty Museum

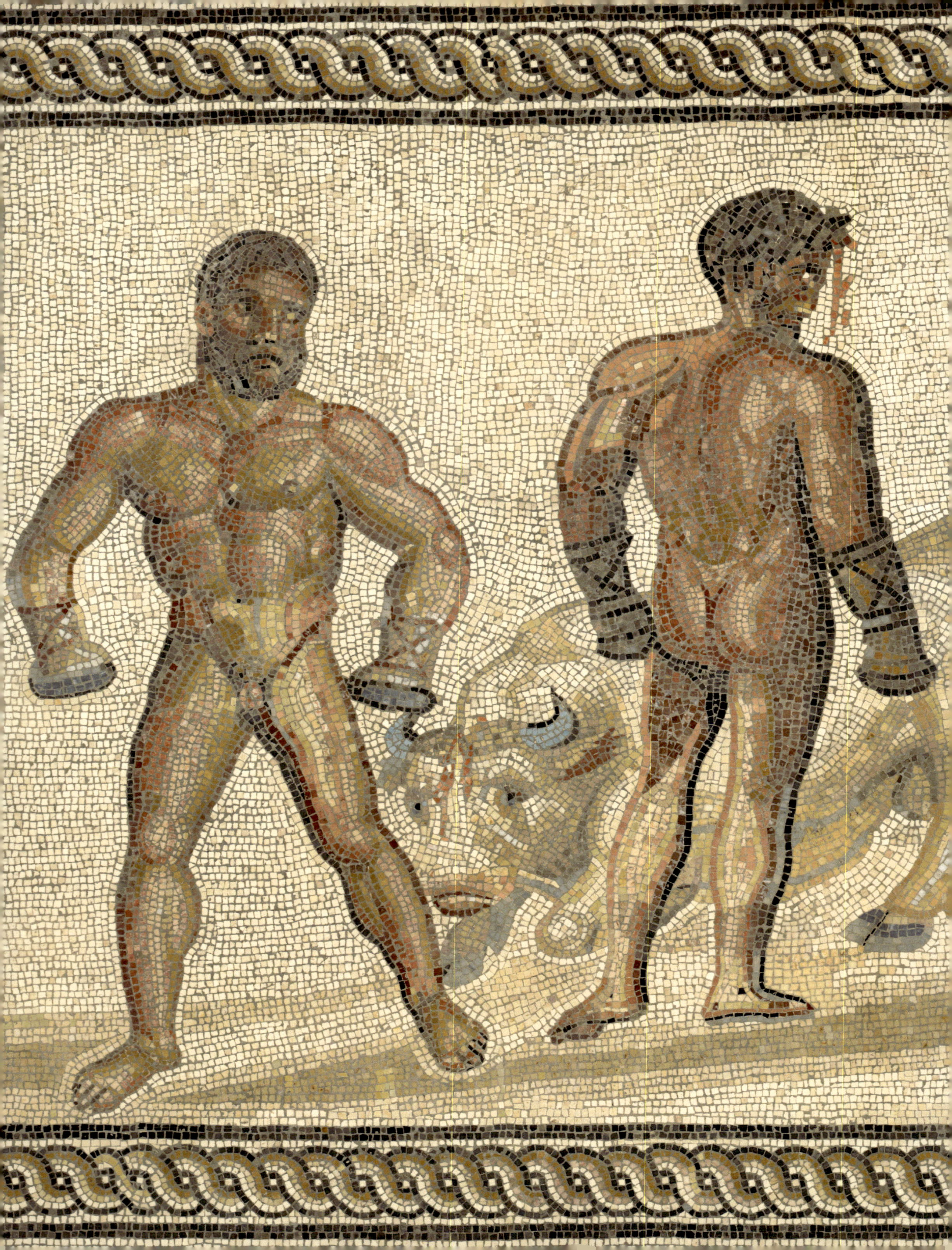

Introduction
Christine Kondoleon

The mosaics in the collection of the J. Paul Getty Museum span the second through the sixth centuries AD and reflect the diversity of compositions found throughout the Roman Empire during this period. Several of the mosaics in the Getty collection can be traced to specific discoveries made in the nineteenth and early twentieth centuries in Italy, Gaul, North Africa, and Syria. Although the Greeks introduced the medium, using pebbles embedded in mortar, it was really the Romans, with their use of tesserae in a technique called *opus tessellatum*, who fully explored it as suitable ornamentation for their architecture. In fact, mosaic floors are closely aligned with Roman culture, and because of their durability they have survived in greater numbers than paintings and sculptures and testify to a lively and imaginative practice of decorative and figurative arts.

The placement of mosaics can suggest the function of the spaces they once decorated, and certain mosaic themes call on the visitor to interact with the work in a specific and intentional way. Unlike wall paintings, floor mosaics insist on movement over them and engage the viewer in a surprisingly physical experience. In order to better understand these compositions, it is critical to keep their spatial and tactile qualities in mind. Through the hundreds of visual images that survive in Roman mosaics, we can sometimes read the aspirations, anxieties, and pleasures of those who lived in the houses, towns, and cities of the Roman Empire.

The Roman architect Vitruvius, in his treatise on architecture, *De Architectura*, documents the techniques used for the preparation of mosaic pavements, but he does not indicate how the tesserae were gathered or cut, or how the compositions were laid out. The use of mosaic pavements followed the spread of Roman culture throughout the empire during the Imperial period (about the first through fourth centuries AD). When painterly effects were desired, the colored stones were cut into very small tesserae, and sometimes glass was used to create tones of orange, yellow, blue, and green not easily found in nature. In view of the grand pictorial works such as the Alexander Mosaic from Pompeii and the number of similar examples found throughout the empire, it is reasonable to assume that mosaicists, like painters and weavers, relied on cartoons, drawings, and sketchbooks. However, the artisans were highly skilled and clearly at liberty to be flexible and imaginative with their commissions, so no one mosaic is exactly like another.

As the empire expanded, the elite increasingly displayed their wealth and Roman identity in private settings. Regional workshops developed to meet the increasing demand for domestic mosaics. The mosaics in the Getty's collection reveal a rich variety of approaches from different parts of the Roman Empire. Certain themes and organizational patterns were typical of particular regions. In Italy, for example, black-and-white style mosaics came into fashion in the late Republic (mid-second to first century BC) and remained popular well into the third century AD. The Getty Medusa (cat. 1), most likely of the Hadrianic period (during the reign of the emperor Hadrian from AD 117 to 138), is largely black and white with a polychrome center featuring a frightening serpentine visage. Triangles of black and white produce a swirling circular pattern around the Gorgon's head, directing the focus to the central image, which in this case was probably also the center of the room. The fact that almost all the extant Medusa mosaics—about one hundred—are set within a kaleidoscopic pattern that produces the impression of motion suggests that the ancients believed the kinesthetic effect of ornament could work in tandem with the mythic image of the Gorgon to ward off evil.

In the Gallic provinces of Rome, mosaic compositions are largely arranged in a field of patterns divided into compartments spread evenly across the floor. This approach is well illustrated in a mosaic from Saint-Romain-en-Gal in southern France (cat. 3). The very popular subject of Orpheus charming the beasts is set out within hexagonal compartments inscribed within a circle, demonstrating how geometry and ornament were equally balanced with figural subjects in this region. Two other Gallic mosaics come from a Roman villa at Villelaure, near Aix-en-Provence. Multiple borders, including the braiding and chevrons typical of the Roman workshops in Gaul, surround each part of the floor's designs. The larger panel (now in the Los Angeles County Museum of Art; see fig. 14), with Diana as huntress at its center, is surrounded by four different hunting scenes. These four vignettes face in different directions, offering multiple viewpoints to those walking over the mosaic and suggesting that it once decorated a reception space. The pendant panel at the Getty (cat. 4) features a scene from

Virgil's *Aeneid*, in which the Trojan Dares fights a local Sicilian named Entellus. While the figural elements of pendants typically relate to one another thematically, it was not always so, and in this case there may be only a broad connection—perhaps the two mosaics represent ideal masculine endeavors.

The mosaics of Roman North Africa are quite different from those of Italy and Gaul in their exuberant embrace of color and scale; large-scale figural mosaics spread out across the floors, accommodating rooms of many shapes and sizes. However, the very first mosaics in this region, dating to the late first and second centuries AD, rely on Hellenistic Greek traditions. This is well demonstrated by a mosaic showing a lion attacking a wild ass, or onager (cat. 5), which is set in a rocky landscape beside a pool of water, purportedly from Hadrumetum (present-day Sousse in Tunisia). This mosaic was created as a separate picture panel known as an *emblema* and made of tiny pieces of stone and glass arranged to imitate painting. The subject and the style recall the famous panel from Hadrian's Villa in Tivoli (now in Berlin), which depicts a battle of centaurs and wild beasts in a rocky landscape. A similar theme of wild felines attacking onagers occurs in two panels from a dining room in the House of the Dionysiac Procession in Thysdrus (present-day El Djem in Tunisia), but these panels are probably of a later date than the Getty mosaic, as there is a less pronounced illusionistic handling of the landscape.

The Hellenistic tradition of *emblemata* remains strongest in the Greek-speaking part of the empire, namely the Roman provinces of Arabia, Cilicia, Cyprus, Palestine, and Syria. Many floor mosaics found in Antioch, the ancient capital of Syria, show a preference for the framed picture panel associated with the Hellenistic style of earlier periods. The Getty's Achilles and Briseis panel (cat. 6) may be closely related to two mosaics from Antioch, one from the second century and the other from the third century AD, which illustrate the same episode from the *Iliad*. Most likely the Getty panel is also from a Syrian workshop but presents a more detailed composition than either of the Antiochene examples. The painterly effects are obvious in all three mosaics despite their fragmentary states, and they underline how committed the Eastern artisans were to depicting paintings in stone.

The stylistic interconnections of these regional mosaic workshops suggest that many artisans were itinerant. The North African school, once established, had a far-reaching influence. We can see this in the example of the Bear Hunt mosaic (cat. 2). The scale and style of the Getty mosaic evoke that of a large multicolored mosaic found on the Esquiline Hill in Rome (near the Church of Santa Bibiana), which was possibly part of the Constantinian imperial palace known as the Sessorium. (Today, the mosaic is on view at the Musei Capitolini Centrale Montemartini.) In this mosaic, hunting scenes from a stock repertory depicting the capturing of the wild animals for the *venationes* (animal hunts) are distributed across a wide rectangular panel, which perhaps once covered a long palace hall. Both of the fourth-century AD Italian mosaics exhibit themes that can be directly linked to the gathering of the beasts for the *venationes* featured on the great portico of the Villa of Piazza Armerina in central Sicily, which also dates to the fourth century. There is little doubt that North African workshops took part in the massive commission of the Villa of Piazza Armerina. Their work stands out for the exceptional realism of the staged wild beast hunts that began to appear in their repertory in the late third century AD.

Mosaics featuring hunts and wild beasts have been found throughout the Mediterranean and have proven to be one of the most popular themes in the Greek East as well as in the Latin West. The display of animals lent itself well to the demand for pavements covering large and irregularly shaped spaces, as these subjects could be scattered about, facing in different directions. This is the case with a group of fragments (depicting a griffin, a horse, a lion, bulls, a rabbit, a donkey, a stag, birds and a tree, and two peacocks; see cats. 8, 9–19) whose style, subject, and arrangement clearly belong to an eastern Mediterranean workshop of the fifth or sixth century AD. The types of subjects, especially the peacocks and the vines and their free distribution within the compositions, are characteristic of the treatment of a number of nave mosaics in Christian churches in the Syrian region at this time. This shift from the iconography of the Roman amphitheater to the realm of Christ demands an explanation. By the early Byzantine period, the beasts in such parades are often represented as quite tame and are thought to refer to the peaceable kingdom of paradise, and may even more broadly imply the domain of the Lord and his Creation. The boundaries between secular and religious imagery were quite permeable in the new Christian Empire.

ITALY

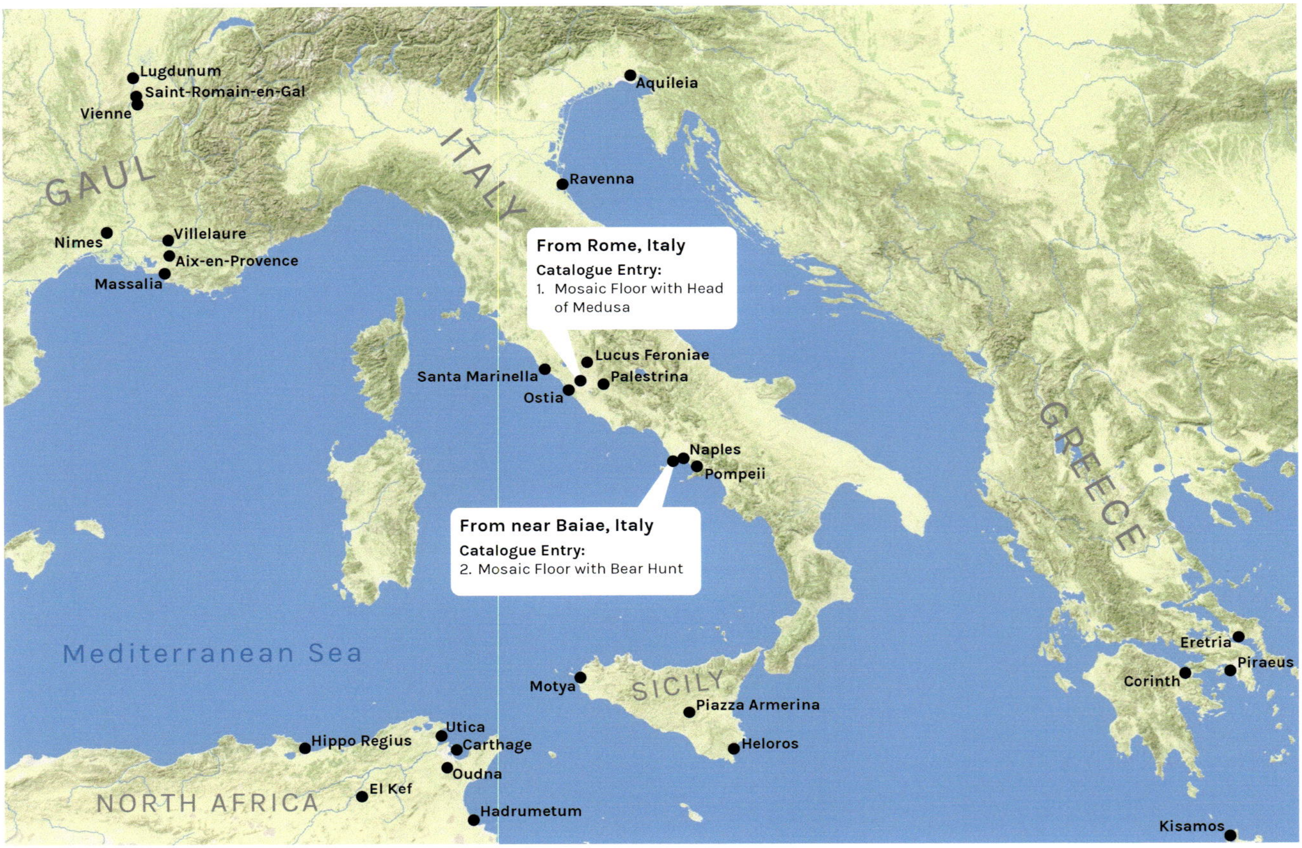

Italy

The mosaics of Italy first appeared in the late second century BC under the influence of the Hellenistic Greek pictorial tradition, in which tiny pieces of irregular stone were used to create narrative themes in detailed, colorful compositions that imitated the effects of painting.[1] The earliest examples of this type, most of which were *emblemata*, or picture panels, were either introduced to the region by Greek craftsmen or manufactured elsewhere and imported. Found predominantly in the cities around the Bay of Naples and Rome, mosaics in this style, such as the well-known Alexander Mosaic from the House of the Faun at Pompeii and the Nile Mosaic from Palestrina, both dating to about 100 BC, were considered expensive luxury items, reserved exclusively for the decoration of the wealthiest constructions. Smaller *emblemata* were usually centerpieces of floors, and they were typically surrounded by plain white tesserae or framed by simple decorative patterns. On some later floors, the figural scene was set in a more elaborate design, such as in the House of the Labyrinth at Pompeii, which features an *emblema* of Theseus and the Minotaur dating to about 70–60 BC and framed by a black-and-white labyrinth pattern.[2]

By the end of the first century BC, the production of mosaic pavements in various new Roman styles became widespread throughout Italy. The tremendous expansion in the use of mosaics to decorate both public and private buildings resulted in a number of different types that ranged from colorful figural scenes to predominantly black-and-white geometric and floral patterns.[3] The Villa of the Volusii Saturnini at Lucus Feroniae, north of Rome, includes both styles: polychrome mosaics in a phase of the villa dated to about 60–50 BC contrasting with black-and-white pavements from the period of about AD 10–20.[4] Black-and-white style mosaics, like the Getty's Mosaic Floor with Head of Medusa (cat. 1), however, became increasingly common by the late first century AD, especially in Rome and central Italy.[5] Complex geometric designs, like the intricate pattern of curvilinear triangles encircling the bust of Medusa in this mosaic, now covered the floors of entire rooms. Stylized vegetal motifs, used in similarly themed compositions, such as the detailed floral pattern that surrounds the central head of Medusa in a mosaic decorating the floor of the *triclinium* in the House of Bacchus and Ariadne at Ostia, were equally prevalent.[6] Although this new decorative style may have served as a less expensive solution to the more costly tesserae required for polychrome mosaics, in the second century AD black-and-white patterned mosaics of extremely high quality were also used in the decoration of Hadrian's Villa at Tivoli and in the Palazzo Imperiale at Ostia.[7]

The growing popularity of the black-and-white mosaic style in Italy in the first and second centuries AD had a strong impact on mosaic designs outside of Italy, particularly in the Roman provinces of Gaul and North Africa. The introduction and diffusion of this style in these regions in the first century AD can most likely be attributed to the increasing Roman presence in the region; itinerant craftsmen traveled or settled in the area and established workshops for mosaic production. Around the beginning of the fourth century AD, however, provincial workshops exerted a profound influence in Italy, even in Rome itself, bringing a new preference for large-scale scenes of the hunt or the arena to luxurious Italian villas. The Getty's fourth-century AD Mosaic Floor with Bear Hunt (cat. 2), from a seaside villa near Baiae on the Bay of Naples, exemplifies this new style, which was particularly characteristic of mosaics in North Africa during this period. The North African influence is also evident in the Great Hunt mosaic at Piazza Armerina in Sicily and in a hunting mosaic from the Gardens of Licinius on the Esquiline Hill in Rome, both dating to the early fourth century AD, which in turn derive from the third-century AD hunting mosaics, such as those from Carthage, Thysdrus (present-day El Djem), and the Villa of the Laberii at Oudna.[8]

NOTES

1. This technique, known as *opus vermiculatum*, appeared as early as the third century BC in Sicily and then spread north throughout Italy. On the origins of these types of pavements, see the discussion in Phillips 1960, 243–62. For a general history of mosaics in Sicily, see von Boeselager 1983.
2. Strocka 1991.
3. The earliest black-and-white ornamental mosaics often imitated other types of pavements in motif and composition, and they may have been used as a substitute for the more expensive polychrome mosaics.
4. Moretti and Sgubini Moretti 1977, plates 35–36, 39–42. The new preferences that begin to appear in mosaics about 20 BC also parallel changes in taste in wall painting: polychromy and three-dimensional decorative motifs disappear and are replaced by flat black-and-white designs and figural styles; see Clarke 1991, 61–63.
5. On the relationship between the development of black-and-white style mosaics and the architectural spaces they decorated, see Clarke 1979.
6. House of Bacchus and Ariadne: Clarke 1979, fig. 20.
7. Hadrian's Villa at Tivoli: Blake 1936, plates 11–14. Palazzo Imperiale at Ostia: Becatti 1961, plates 24, no. 300, and 69, no. 296. Ostia, the port of Rome, provides the most extensive evidence for the development of mosaic pavement types following the destruction of Pompeii and Herculaneum by the eruption of Vesuvius in AD 79; many examples date to the second and early third centuries AD. The vast majority of these are black-and-white figural mosaics, often with marine subjects, as in the Baths of Neptune (Clarke 1979, figs. 31–34) and the Baths of the Lighthouse (Clarke 1979, figs. 69–71).
8. Piazza Armerina: Gentili 1954, 33, fig. 14 (Small Hunt), 36, figs. 17–21 (Corridor of the Hunt); Gentili 1959, figs. 4, 5 (drawings); and Lavin 1963, 244–51, figs. 8, 9 (Corridor of the Hunt), fig. 110 (Small Hunt). Esquiline: Aymard 1937, 42–66, plates 1–3; Lavin 1963, 258, figs. 122–23; and Cima 1998, fig. 9. The Esquiline mosaic is now in the Musei Capitolini Centrale Montemartini (inv. no. AntCom03636). Thysdrus: Gauckler 1910, 26, no. 64; and Lavin 1963, fig. 80. Carthage: Poinssot and Lantier 1923, 154–58; and Lavin 1963, fig. 79. Oudna: Gauckler 1910, 122–23, no. 362; and Lavin 1963, 230–31, fig. 75.

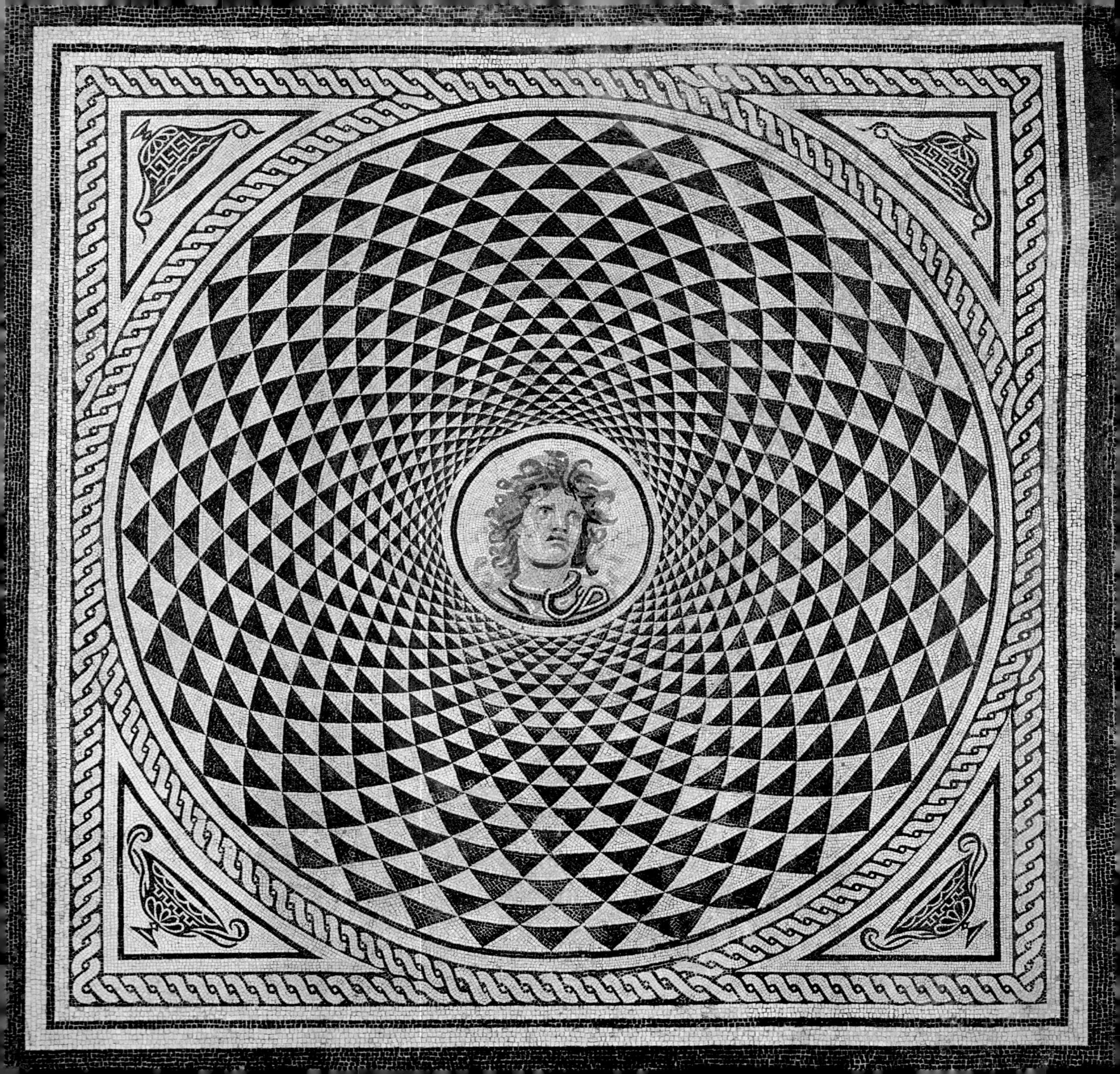

1. Mosaic Floor with Head of Medusa

Roman, from Rome, Italy, AD 115–150
Stone tesserae, 270.5 cm × 270.5 cm
71.AH.110

Provenance

This mosaic originally decorated the floor of a Roman villa in one of two adjacent rooms, both with similar mosaics, discovered on the Via Emanuele Filiberto in Rome. The site was excavated by Angelo Pasqui in 1910, but it is uncertain when the mosaic was removed.[1] It was purchased by the J. Paul Getty Museum in 1971.[2]

Commentary

A colorful female bust is the only polychrome feature in this otherwise black-and-white mosaic. Snakes around her neck and serpentine locks of hair identify her as the Gorgon Medusa. The image decorates the central medallion of the mosaic, with the face turned upward and to the right. An elaborate geometric pattern encircles the bust: concentric bands of alternating black-and-white triangles decrease in size as they spiral toward the center, creating the optical illusion of spinning motion. The design has also been interpreted as a shield of scales, a reference to the aegis worn by Athena, a scaly mantle decorated with the head of Medusa. A *guilloche* border around the circle is enclosed within a second, square *guilloche* that outlines the entire composition. *Kantharoi* fill the triangular spaces at each of the four corners.

In Greek mythology, the Gorgon Medusa was a fearsome monster who turned viewers to stone with her gaze. When she was finally killed—beheaded by the hero Perseus—her hideous head was presented to his patron goddess, Athena. The Gorgon head was a popular apotropaic device in Greek art, as its terrifying appearance was believed to ward off evil. In Roman art, however, Medusa was humanized and more clearly female; at times she was even depicted in the form of a beautiful woman. During the Roman period, the image of the Gorgon often served a primarily decorative function in interior decoration, appearing, for example, on domestic utensils and wall paintings, but it continued to be regarded as a protective symbol. Representations of Medusa were often accompanied by imagery related to the god of wine, Dionysos, whose worship invoked pleasure and good fortune. The *kantharoi* found in the corners of the Getty mosaic are closely associated with the revelries of Dionysos. A Roman mosaic from Kisamos, on Crete, similarly depicts the central bust of Medusa surrounded by panels of masks and followers of Dionysos.[3] An explicit connection appears in a late second-century AD mosaic pavement from a Roman villa at Corinth, which features the head of Dionysos at the center of the same pattern found on the Getty mosaic and a *guilloche* border with *kantharoi* in the corners.[4]

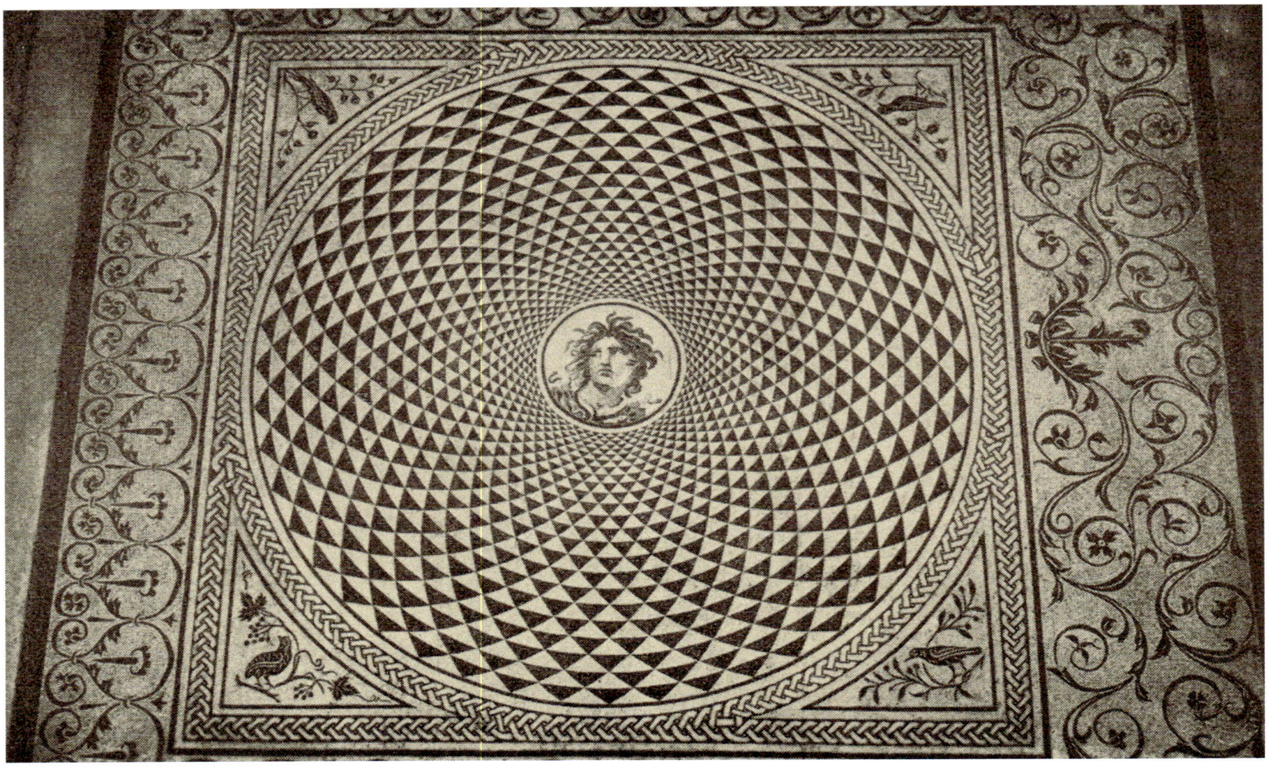

Figure 1. Mosaic with Head of Medusa, Roman, AD 115–150. Found in Rome, Italy, 1910. Stone tesserae, 530 × 450 cm. Museo Nazionale Romano—Palazzo Massimo alle Terme, 56253

The Getty's Medusa mosaic originally decorated a small room measuring 3.6 × 2.9 meters. A pendant mosaic (fig. 1) survives from an adjacent, larger room (5.3 × 4.5 meters) of the same villa; it was most likely made by the same workshop and is now in the collection of the Museo Nazionale Romano—Palazzo Massimo alle Terme.[5] This second mosaic depicts the bust of Medusa in mirror image, her gaze turned upward to the left. Her head is also placed at the center of a black-and-white shield or spinning wheel design. Birds perch on branches in the four corners instead of *kantharoi*, and a pattern of stylized tendrils and ornate scrolls frames the *guilloche* border. The excavation of this villa revealed that the two rooms originally formed one large rectangular chamber decorated with a black-and-white mosaic of a marine scene before it was subdivided and redecorated with the pair of Medusa mosaics.[6] Although the reason for the renovation or reconstruction of the villa is unknown, the construction of the building in *opus mixtum*, a technique used especially during the rule of the emperor Hadrian from AD 117 to 138, indicates that the transformation likely took place sometime in the early second century AD.[7] The style of the mosaics is typical of private villas in Italy during this period, which were decorated predominately with black-and-white mosaics made up of complicated, often curvilinear geometric patterns and highly stylized floral designs.

Comparanda

The circular geometric design with the Gorgon head at its center was used frequently on mosaic pavements throughout the Roman Empire. Many were colorful, polychrome compositions. In second-century AD mosaics from Piraeus in Greece and Pergamon in Asia Minor (present-day Turkey), the triangles of the scale pattern are rendered in shades of red, blue, green, and yellow, with a *guilloche* border in yellow and blue.[8] The example from Pergamon, like the Getty mosaic, also displays *kantharoi* in its four corners. In a variation on the spinning aspect of the design, the curvilinear triangles encircling the head of Medusa in a late second-century AD mosaic floor from the House of the Red Pavement at Antioch form concentric petal shapes.[9] Scale patterns could also be composed of shapes other than triangles. A mosaic from the *triclinium* of a Roman villa at Marcianopolis in Thrace (present-day Devnya, in Bulgaria), known as the House of Antiope, displays Medusa within a circular pattern of spade-shaped scales surrounded by a meander pattern, with felines in each of the corners.[10] Comparable examples from North

Africa include a Roman villa at Thysdrus (present-day El Djem) and a bath complex at Dar Zmela, both characterized by a similar pattern of spade-shaped scales and flower motifs in the corners, as well as the Great Baths at Thaenae (present-day Thina), decorated with radiating waves of alternating colors.[11] An unusual version of this design in *opus sectile* was discovered at Kibyra in southern Turkey, where it covered the floor of the orchestra of an *odeion* or *bouleuterion* that was destroyed in the fifth or sixth century AD.[12] Although other examples of the shield or spinning wheel motif appear in *opus sectile*, all feature a geometric pattern in place of the head of Medusa. One of these, executed in polychrome marble, was found in the Villa dei Papiri at Herculaneum, the famous Roman villa that inspired the design and decoration of the Getty Villa in Malibu.[13]

Condition

The central medallion is largely intact, but the edges, including the two vessels in the lower half of the mosaic and much of the border, appear to be modern reconstructions. Reportedly, the mosaic was already badly damaged when it was discovered.

Bibliography

Pasqui 1911, 338–39; Blake 1936, 83; Vermeule and Neuerburg 1973, 51–52, no. 111; Hessenbruch McKeon 1983, 239–41, cat. no. 23; Aisenberg 1994, plate 10; J. Paul Getty Museum 2010, 220.

A.B.

NOTES

1. For the excavations of the villa, see Pasqui 1911, 338–39. It seems that the mosaic may have been reburied and lifted later. Period photographs in the Getty curatorial files, including one labeled "Casa dei Neroni," show the mosaic as the floor of a modern house, but the location and previous owner have not yet been identified.
2. Purchased from the Royal Athena Galleries, New York.
3. Markoulaki 1987, plate 324 (detail only), early third century AD. Another Medusa mosaic, from the Villa of Dionysos at Knossos on Crete, depicts heads representing the four seasons in the corners; see Paton 2000, 553–62.
4. Shear 1930, plate 10 (now in the Archaeological Museum of Ancient Corinth, inv. no. A609 / MOS 25–3).
5. Museo Nazionale Romano, Rome, inv. no. 56253; Paribeni 1913, plate 1; and Aurigemma 1974, plate 104. In Hessenbruch McKeon 1983, 238–39, cat. no. 22, the author writes that at the time of publication the tondo was in the Ministero Archeologico di Roma. It is currently in the Museo Nazionale Romano—Palazzo Massimo alle Terme. No architectural plan or description of the layout of the building was published at the time the mosaic was excavated.
6. See Pasqui 1911, 338, with figures of sea monsters and dolphins led by winged putti and Tritons.
7. See Pasqui 1911, 338, which identifies the remains of a building made in *opus mixtum*.
8. Piraeus (now in the National Archaeological Museum of Athens): Philadelpheus 1894, plate 4. Pergamon (now in the Bergama Archaeological Museum, Izmir): Scheibelreiter-Gail 2011, 338–39, no. 107, and 649, fig. 484. Additional information can be found in Hessenbruch McKeon 1983, 264–67, nos. 43 and 45.
9. Levi 1947, plate 14a (now in the Princeton University Art Museum, inv. no. y1965-212).
10. Minchev 2002, 253, figs. 4, 5. The mosaic remains in situ in the Museum of Mosaics, Devnya (Bulgaria), which was constructed on top of a late third- to early fourth-century AD Roman villa, the so-called House of Antiope.
11. El Djem: Merlin 1915, no. 71f, no. 2; Foucher 1963, 97, fig. 13b; Dunbabin 1978, 163 and 258 (catalogue). Dar Zmela: Foucher 1960b, 121–22, no. 57.247; Foucher 1963, 97, fig. 13a; and Dunbabin 1978, 163 and 271 (catalogue). The design is meant to represent a shield, but it is not contained within a circular frame. Thina: Gauckler 1910, no. 18, A, 14; and Dunbabin 1978, 163 and 273 (catalogue). Additional bibliography can be found in Hessenbruch McKeon 1983, 301–2, nos. 64, 71, 72. Dunbabin 1978 dates the mosaics from El Djem and Thina to the late third century AD and the mosaic from Dar Zmela to the second half of the second to early third century AD, identifying Medusa as an *apotropaion*.
12. Özüdoğru and Dökü 2010, 39–42, figs. 5, 6; and Özüdoğru and Dökü 2012, 51.
13. Spinazzola 1928, plate 195, now in the Museo Archeologico Nazionale, Naples.

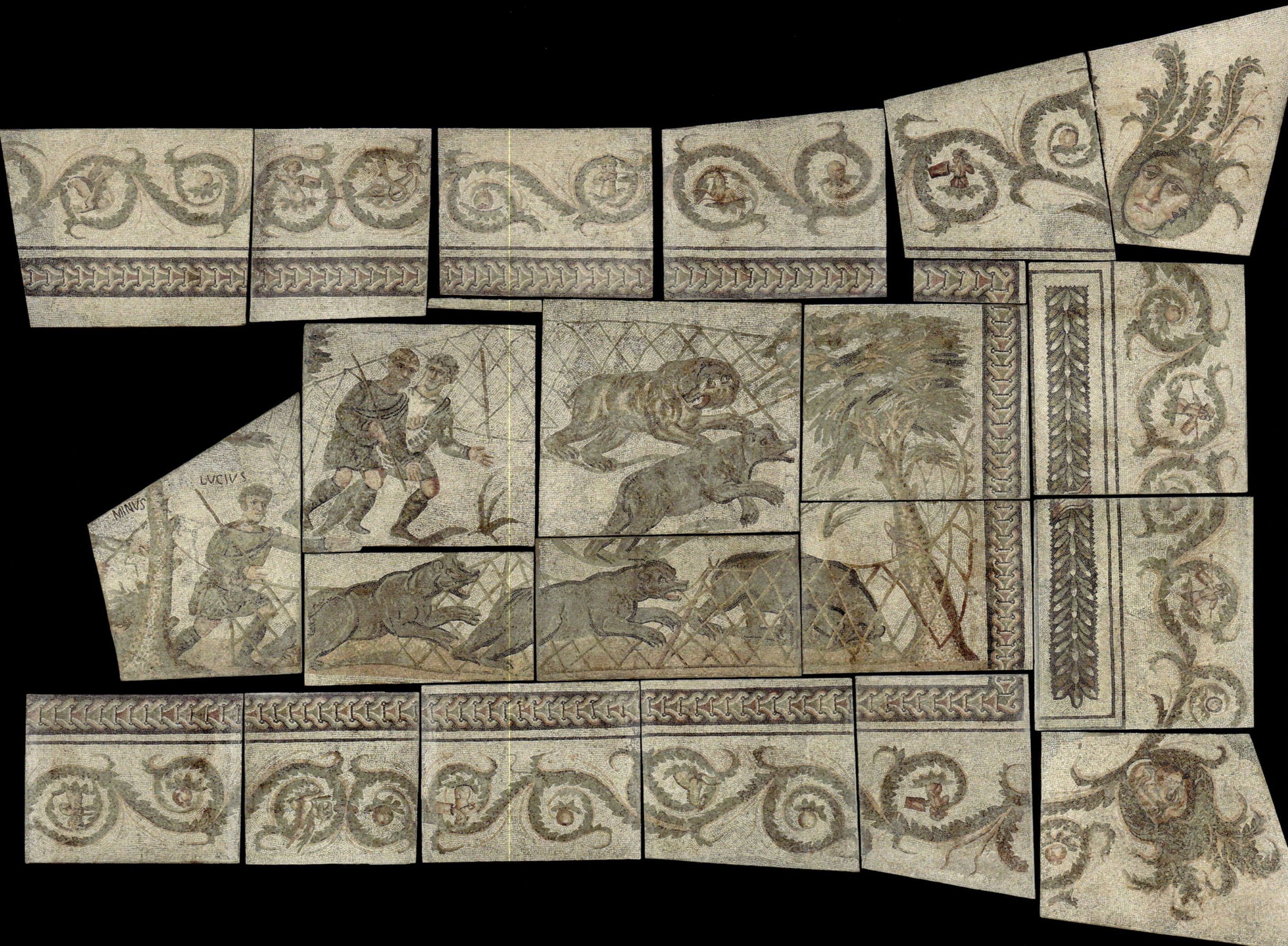
LVCIVS
MINVS

2. Mosaic Floor with Bear Hunt

Roman, from near Baiae, Italy, AD 300–400
Stone tesserae, 661 cm × 869 cm
72.AH.76.1–.23

Provenance

This mosaic was acquired by the J. Paul Getty Museum in 1972 from Jeannette Brun, a Zurich-based antiquities dealer, who reported that it had been in an Italian collection but provided no further information regarding its provenance. Subsequent publications attributed the work on stylistic grounds to a North African atelier.[1] Recent archival research, however, indicates that the mosaic was in fact unearthed in June 1901, in a vineyard in the vicinity of Lago di Lucrino, north of Baiae and just west of Naples.[2] At the time of its discovery, architects and archaeologists at the Museo Archeologico Nazionale in Naples suggested that the mosaic may have decorated the great room of a bath but believed it had little artistic value and recommended against its acquisition by the museum.[3] Subsequently, the Italian state authorized the landowner, Schiano Muriello, to sell the mosaic, and it was purchased in March 1906 by Ernesto Osta, a lawyer who intended to use it in the decoration of the monument to King Vittorio Emanuele II in Rome. The mosaic was lifted from the ground, but portions of it remained in situ owing to its poor condition and the difficulty of removing it intact. The lifted mosaic remained in storage in Naples, and in 1925 it was offered to John Marshall, a British art expert and dealer in Rome. By 1929, Osta's heirs sold it to Rodolfo Follis of Turin. Follis sought ministerial permission to export the mosaic, but the legality of his ownership was questioned.[4] Precisely when the mosaic left Italy is unknown. Four other panels from the mosaic (figs. 2–5), which had been clandestinely removed from the site and eventually recovered by Italian authorities, are now in the Museo Archeologico Nazionale in Naples.

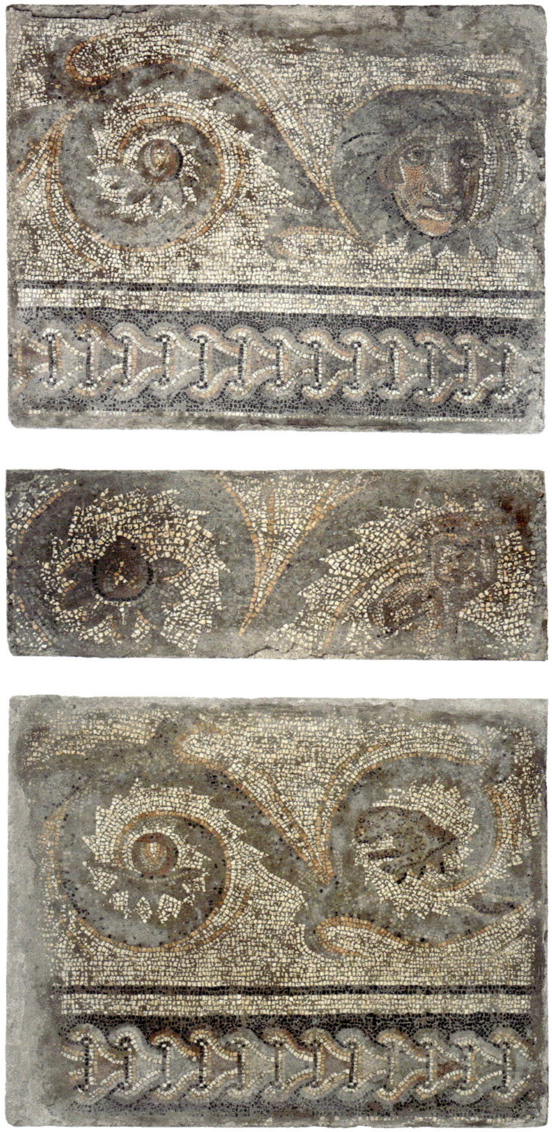

Figures 2–4. Panels from Mosaic Floor with Bear Hunt, Roman, fourth century AD. Found in Baiae, Italy, 1901. Stone tesserae, 115 × 141 cm (top), 69 × 156 cm (center), 116 × 142 cm (bottom). Museo Archeologico Nazionale di Napoli, MANN 11474–76

Commentary

The center of the preserved portion of the mosaic depicts a bear hunt. From the left, three hunters wearing high boots, long-sleeved tunics, and short, belted garments draped diagonally from the left shoulder drive five bears into a large, semicircular net tied to a pair of trees. A fourth hunter appears on a panel now in the Museo Archeologico Nazionale in Naples (fig. 5).[5] His outstretched left arm continues onto the left side of the Getty's part of the mosaic (fig. 6). The others stand in similar poses, with legs bent and arms extended forward. Three hold staffs in their right hands; three are beardless. Two of the figures are accompanied by inscriptions identifying them as Lucius and Minus. A cord in Minus's left hand suggests that he is tying the net to a tree. The bears, like the hunters, move to the right across different levels of ground. All the bears have open mouths, and two have turned their heads back to the hunters, as if to snarl at them. In addition to the two tall trees that frame the scene, other, abbreviated landscape elements include darker tesserae representing the ground in the lower level and some small plants and clumps of grass above. Cursory shadows are also depicted, as are highlights and shading.

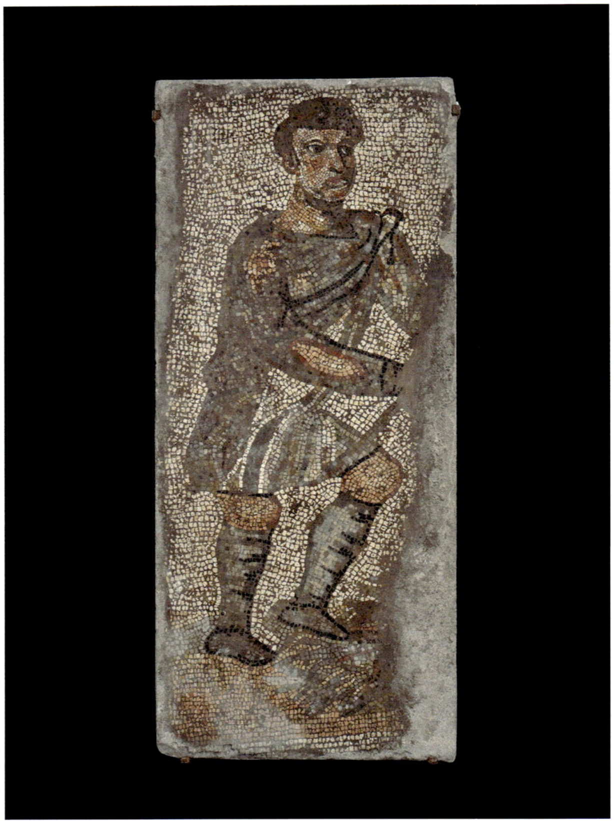

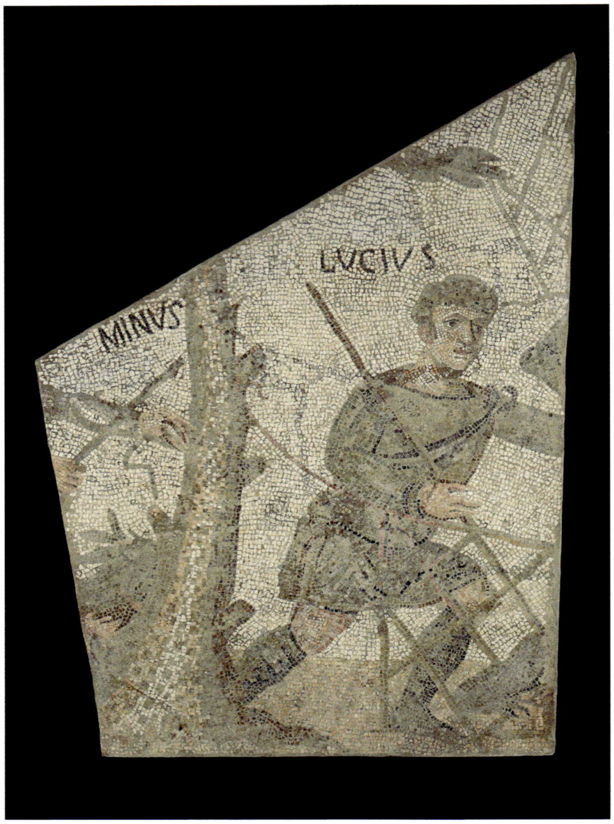

Figure 5. Panel from Mosaic Floor with Bear Hunt, Roman, fourth century AD. Found in Baiae, Italy, 1901. Stone tesserae, 112 × 40 cm. Museo Archeologico Nazionale di Napoli, MANN 11477

Figure 6. Panel from Mosaic Floor with Bear Hunt, Roman, fourth century AD. Found in Baiae, Italy, 1901. Stone tesserae, 194 × 142 cm. Malibu, J. Paul Getty Museum, Villa Collection, 72.AH.76.7

The figural scene is bordered by a colorful double *guilloche* consisting of green, yellow, white, red, and black tesserae. Outside the *guilloche*, at right, is a long, straight, vertical laurel festoon tied at center by a ribbon. A third, outermost border consists of an exuberant acanthus rinceau inhabited by fruit, armed cupids, and the protomes of real and imaginary animals, including a horse, a panther, and a griffin. The two preserved corners are adorned with large acanthus-enveloped faces (fig. 7), rather like personifications of the Seasons but undifferentiated. On one of the four panels in Naples, a third, smaller face emerges from the acanthus rinceau (fig. 2). This face was likely located beneath and to the left of the hunters, and its almost fully frontal orientation suggests that it may have originally marked the midpoint of the entire mosaic, which must have included at least one additional scene; for while the right side of the bear hunt terminates in a tree and the vertical festoon beyond the double *guilloche* border, the left side evidently continued beyond the tree between Minus and Lucius. Other large hunt mosaics of the period, such as that from the Gardens of Licinius on the Esquiline Hill in Rome, also include the netting of animals as part of more complex compositions consisting of several different scenes.[6]

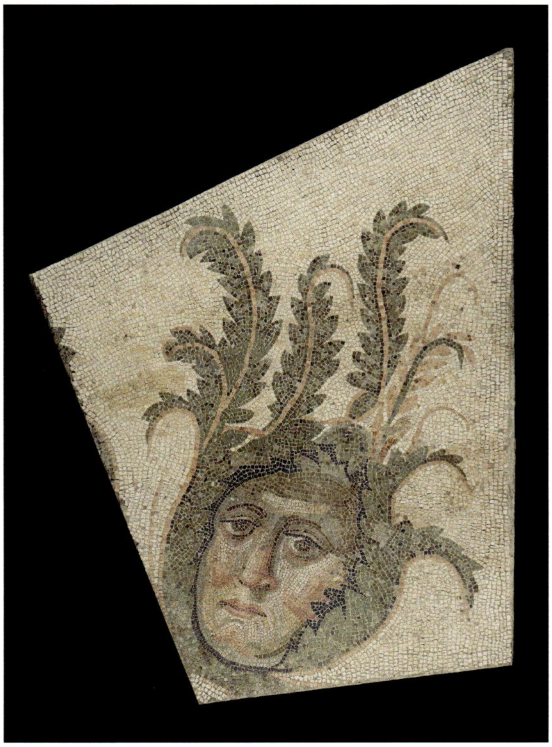

Figure 7. Panel from Mosaic Floor with Bear Hunt, Roman, fourth century AD. Found in Baiae, Italy, 1901. Stone tesserae, 174 × 148 cm. Malibu, J. Paul Getty Museum, Villa Collection, 72.AH.76.6

Numerous ancient literary sources note the aristocratic taste for *venationes*. The emperor Hadrian, who ruled from AD 117 to 138, is reported to have hunted bears in both Greece and Asia Minor, and he is depicted hunting on horseback in a roundel on the Arch of Constantine in Rome. The emperor even composed a poem, later inscribed in stone, celebrating his success as a hunter. Bears were found throughout the ancient Mediterranean world, however, and were captured to be trained as well as slaughtered. The second-century AD treatise on hunting, *Cynegetica*, by Pseudo-Oppian, describes in detail one method of netting bears (although not the manner depicted in the Getty mosaic): driving them along a rope hung with colored ribbons and feathers. The letters of the late fourth-century AD consul Symmachus, among other sources, relate efforts to procure bears and various beasts for public entertainment.[7] Klaus Werner, who first recognized the Campanian origins of the Getty mosaic, and Maddalena Cima have argued that this theme was a particular favorite of emperors who sponsored spectacles with wild animals in the amphitheater.[8]

The findspot of the Getty Bear Hunt mosaic has yet to be fully investigated. It was discovered with a marble border and fragments of columns near the so-called Stufe di Nerone, in the Scalandrone neighborhood north of Baiae. These monumental architectural remains have been damaged by the construction of modern streets and overbuilding and, consequently, are not well understood.[9] The early twentieth-century archaeologists who saw the Getty mosaic in situ made additional soundings and estimated its full length to be at least twelve meters.[10] The unusual shape of the mosaic, with extended corners, suggests that it may have occupied a space between two more or less oval rooms, such as those used for baths or audience halls.[11] The determination that this mosaic once decorated a rich senatorial villa or an imperial residence must await further investigation.

Comparanda

Scenes of hunters forcing bears and other animals into nets and traps are depicted on mosaics throughout the Roman Empire, including examples from Carthage, El Kef, Hippo Regius, and Utica in North Africa; Centcelles in Spain; Villelaure (see fig. 14) in France; and Rome (including one from the Esquiline Hill, noted above) and Ravenna in Italy.[12] In 1973, Norman Neuerburg suggested a North African origin for the Getty mosaic, although he also noted comparanda for its iconography in mosaics from the city of Rome and elsewhere in Italy.[13] A decade later, David H. Ball published the mosaic, believing it to be of Tunisian origin and attributing it to a Carthaginian workshop. He proposed that it depicted bears being captured for display in the amphitheater, and he was particularly interested in the artists' treatment of space and the possible source of the imagery, which, he believed, derived from illustrated texts of Pseudo-Oppian's *Cynegetica*.[14] At about the same time as Ball's study, Mario Pagano also published the mosaic, not knowing that the majority of it had been acquired by the J. Paul Getty Museum.[15] Working from early reports, archival documents (including a now-lost watercolor), and the four panels in Naples, Pagano compared its composition to that of two mosaics in Sicily—the small hunt mosaic at the imperial villa at Piazza Armerina and another from the Villa del Tellaro, near Heloros—as well as to others in North Africa. He also cited parallels for the populated acanthus rinceau of the border on mosaics in Antioch (Turkey) and Argos (Greece). The signature of the mosaicist T. Senius Felix from Puteoli (present-day Pozzuoli, near Naples) on a large mosaic found at Lillebonne in Gaul, dating to the late third or early fourth century AD, suggests that Puteoli was home to workshops that drew upon designs of North African mosaics to decorate the many splendid villas overlooking the Bay of Naples that belonged to rich senatorial clients.[16]

Condition

The mosaic has been divided into multiple sections, all of which are backed with concrete. Twenty-three panels are in the J. Paul Getty Museum, and four are in the Museo Archeologico Nazionale in Naples. When the Getty panels were first lifted, in 1906 (a date confirmed by scraps of Italian newspaper still attached to two of them), they were backed with concrete reinforced by iron rebar. Correspondence between Jeannette Brun and Getty Museum officials in the early 1970s indicates that restoration of the panels was a condition of the purchase. The restoration was completed in Zurich by mid-June 1972 with considerable difficulty because of the size and number of panels. The concrete backings were reinforced with additional rebar, which in some cases doubled the thickness of the panels. When the mosaic arrived in Los Angeles, museum officials complained to Brun about several aspects of this treatment, especially the lack of alignment among the panels.[17] The panels also seem to have been polished smooth prior to the mosaic's arrival. The mortar between the tesserae of one of the corner panels depicting a face enveloped in acanthus (fig. 7) was subsequently partially removed.

Bibliography

Gabrici 1901; Vermeule and Neuerburg 1973, 53–54, no. 113; Boriello and D'Ambrosio 1979, 44, no. 12, fig. 33; Pagano 1983–84, 179–87, figs. 28–33; Ball 1984; Werner 1994, 293; Cima 1998, 436–38, fig. 8; Lapatin 2014; Pisapia 2014.

K.L.

NOTES

1. Vermeule and Neuerburg 1973, 53; and Ball 1984.
2. Gabrici 1901. See also Boriello and D'Ambrosio 1979, 44; Pagano 1983–84, 179; Werner 1994, 293; Cima 1998, 436–38; and Pisapia 2014.
3. Gabrici 1901; and Pisapia 2014.
4. Pagano 1983–84, 179–80; and Pisapia 2014. The John Marshall Archive, British School at Rome, id.no.642 (Marshall's card file B.IV.21).
5. Museo Archeologico Nazionale, Naples, inv. no. 11477.
6. Lavin 1963; Toynbee 1973; and Cima 1998.
7. *IG* VII 1828; Scriptores Historiae Augustae, *Hadrian* 20.13; Dio Cassius, *Historia Romana* 69.10.2.
8. Werner 1994, 292–93; and Cima 1998, 436–38.
9. Gabrici 1901; Pagano 1983–84, 179n66; and Pisapia 2014, 113–14. No architectural plan or description of the layout of the building was published at the time the mosaic was unearthed.
10. Gabrici 1901.
11. Vermeule and Neuerburg 1973, 54.
12. Lavin 1963; and Toynbee 1973, 93–100.
13. Neuerburg mentions festoons in North Africa, subjects of bear (and other animal) driving in Rome, and peopled scrolls in all regions; see Vermeule and Neuerburg 1973, 53–54.
14. Ball 1984.
15. When Pagano published the mosaic, he believed it had been lost, known only from records. It seems that for a time both scholars were working independently on the mosaic, unaware that it was the same object.
16. Pagano 1983–84, 179–87.
17. Correspondence in the files of the antiquities department of the J. Paul Getty Museum; Lapatin 2014.

GAUL

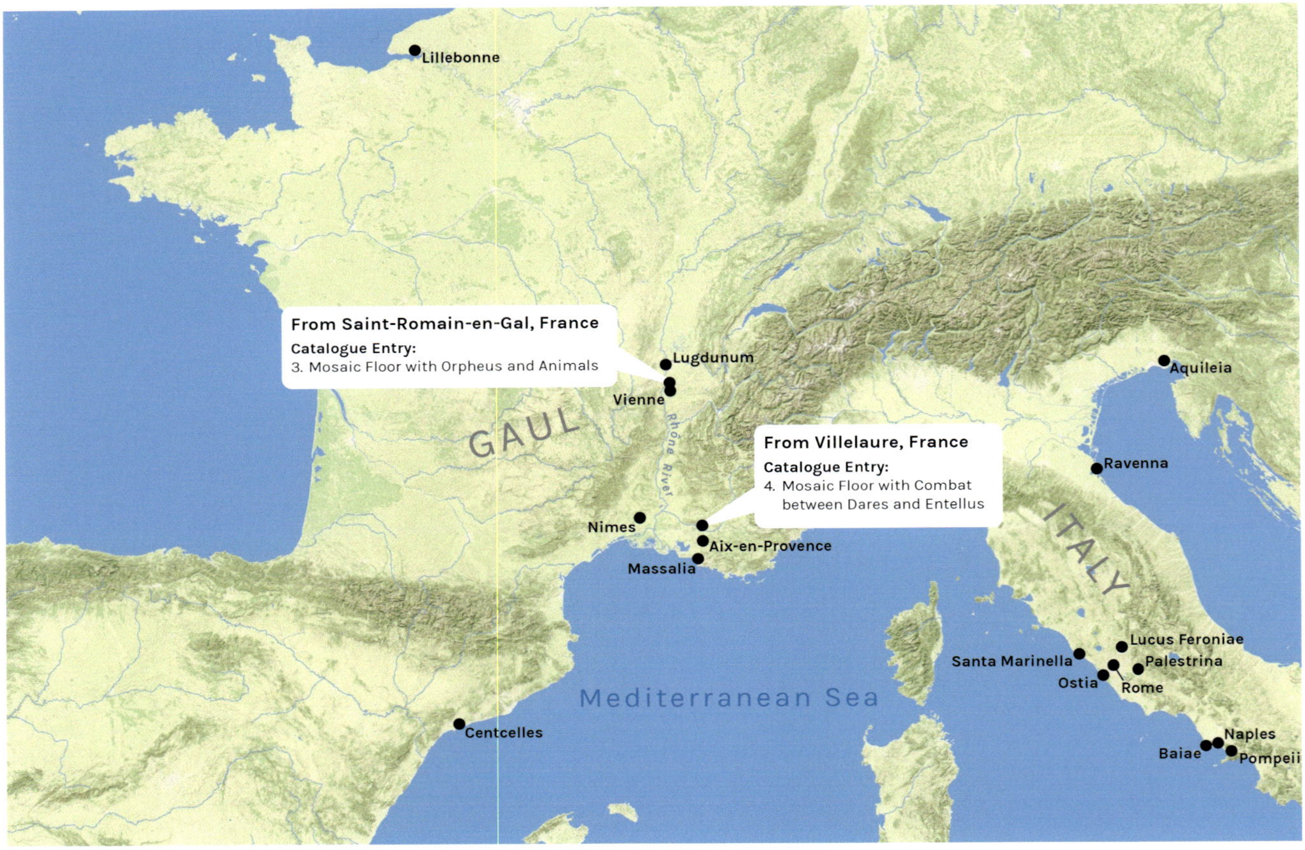

Gaul

The province of Gallia Narbonensis, in the southern part of Gaul (present-day France), was one of the earliest regions of the Roman Empire to be affected by Greek and Roman culture. Even before the Roman conquest of the late second century BC, there was a significant Greek presence in the region, notably at Massalia (present-day Marseille), a Greek colony founded about 600 BC. The earliest examples of mosaics and other decorated floors in Gaul, dating to the first century BC, are found in this region, beginning with simple types related to the Italian *opus signinum* and *terrazzo*. Mosaics composed of tesserae became more prevalent in the late first and early second centuries AD, and it is likely that the technique was introduced by itinerant craftsmen, who established workshops in the region. The development of mosaic styles that followed this early period—specifically, the appearance of the black-and-white style mosaics that were widespread in Italy—reflects the rapidly expanding Roman presence in the region. Local craftsmen soon integrated their own designs and compositions, however, resulting in the distinctive Gallo-Roman style that was characteristic of the workshops operating in the Rhône valley.

The earliest mosaics of this Gallo-Roman type date to the middle of the second century AD. At that time, the main centers of mosaic production in Gaul were located in the upper Rhône valley at Lugdunum (present-day Lyon) and around Vienne, especially in its suburbs of Saint-Romain-en-Gal and Sainte-Colombe, where workshops are thought to have continued production through the early third century AD. These workshops developed the so-called multiple decor design, with individual figures or scenes isolated in an elaborate grid-like framework resembling a coffered ceiling.[1] The *emblema* of the Getty mosaic of Orpheus and the Animals (cat. 3) exemplifies the Gallic multiple decor style, with different figures from the myth of Orpheus inhabiting adjoining hexagons. Notably, however, the surrounding field of black-and-white geometric patterns continues to reflect the influence of Italian traditions (see fig. 8) for the mosaic with its complete border). Another mosaic, also from Vienne, with a central panel depicting the drunkenness of Herakles, displays a more developed form of multiple decor composition. The main scene is surrounded by square panels containing an extraordinary variety of designs that draw upon the extensive decorative repertory of the Rhône workshops.[2] In addition to standard mythological subjects, scenes of contemporary Roman life are also characteristic of mosaics in the upper Rhône valley. A number of mosaics depict episodes from daily life, such as farming and hunting, as well as chariot races and other spectacles from the arena.[3] Representations of these activities were adapted to the multiple decor design by filling in the compartments with figural groups, often acting in a series of related events.

While the main production centers in Gaul were in the upper Rhône valley, an active mosaic culture, dependent primarily on the workshops of Aix-en-Provence, also existed in the cities farther south.[4] The surviving evidence suggests that a more conservative stylistic tradition developed in the south; significantly, the multiple decor style is relatively infrequent there. A mosaic from the rue des Magnans in Aix-en-Provence is a notable exception, in which panels with various geometric patterns surround a scene of the boxing match between Dares and Entellus—the same image depicted on the Getty mosaic (cat. 4).[5] The majority of the Aix-en-Provence mosaics, however, are monochrome pavements consisting of white floors outlined by simple black borders. Although fewer in number, examples of black-and-white geometric designs were also seen in this period, reflecting the Italian influence in this part of Gaul as well.[6] A few mosaics from this area, found almost exclusively in wealthy villas, include colorful central panels with detailed figural scenes.[7]

While many mosaics from the upper Rhône valley depict scenes of Roman life, examples from Aix-en-Provence and its vicinity usually illustrate subjects from mythology or literature. Certain themes used in the mosaics from this area are highly original. The combat between Dares and Entellus depicted on the Getty mosaic is an episode from Virgil's *Aeneid* (5.362–482) that is relatively unknown outside of southern Gaul.[8] Likewise, a second mosaic from the same villa, now in the Los Angeles County Museum of Art, represents a specific scene from Ovid's *Metamorphoses* (2.401–530)—Diana discovering the pregnant Callisto—that has no counterpart in known Roman mosaics.[9] Both mosaics represent distinctly Roman versions of mythology, and their uncommon themes may indicate local preferences associated with this particular area of Gaul.[10]

NOTES

1. For a general discussion of the features of the Gallo-Roman style, see Lancha 1981, 14–16 (with inventory numbers) and a summary in Dunbabin 1999, 74–78. In addition to the multiple decor style, Lancha identifies two additional characteristics: the use of elaborate floral and vegetal elements, and the use of figural scenes and motifs not present in previous Gallic mosaics.
2. Lancha 1981, 106–16, no. 306, plates 40–44, dates the mosaic (now in the Museé de la Civilisation Gallo-Romaine, Lyon) to the last quarter of the second century AD.
3. Lancha 1981, 15–16, lists statistics for common mythological themes, such as Orpheus, Dionysos, Herakles, the Seasons, and the Muses. For examples of farming scenes, see Lancha 1981, 208–25, no. 368, plates 107–23. For chariot races, see Lafaye 1909, nos. 712, 785, 1236. For events from the amphitheater, see Lafaye 1909, nos. 1072, 1295, 1611.
4. Aix has the greatest number of preserved mosaics in Gaul after Lugdunum and Vienne; see Lavagne 2000, 13–14: Aix (129 mosaics), Lyon (156 mosaics), Vienne, Saint-Romain-en-Gal (212 mosaics).
5. Lavagne 1994, 203–15, dated to the mid-second century AD. The pavement lacks anything like the floral designs or the small figural motifs incorporated into examples from Vienne. Regarding the relationship between the rue des Magnans decoration and Vienne, Henri Lavagne asserts that the rue des Magnans mosaic anticipates the multiple decor style that is fully developed in the Vienne workshops; see Lavagne 2000, 17.
6. Lavagne 2000, 13–14, provides fifty-five examples of monochrome pavements and fourteen examples of black-and-white designs. Despite the high proportion of monochrome mosaic pavements in the region, expensive pavements in *opus sectile* were the preferred form of decoration in the numerous luxury Roman villas in this area. Lavagne notes six examples in Aix that, according to him, were made by an Italian workshop, and fifty-five examples total from southeastern Gaul; see Lavagne 2000, 13–14.
7. Lavagne 2000, 13–14, records fourteen examples with figures, all from Aix.
8. Examples of the Dares and Entellus mosaic in Gaul are discussed in cat. 4, in the present volume.
9. Lavagne discusses the scene in the Diana and Callisto mosaic, citing two examples from wall paintings in Pompeii: the House of the Tragic Poet (VI, 8, 3.5) and the House of the Hanging Balcony (VII, 12, 26–27); see Lavagne 2000, 313, no. 916.
10. The mosaics from Villelaure (approx. twenty miles north of the modern town of Aix-en-Provence) were likely produced by the workshops of Aix-en-Provence. The early history of the discovery and excavation of the nearby site is summarized in "Villelaure: History of the Excavations," in the present volume.

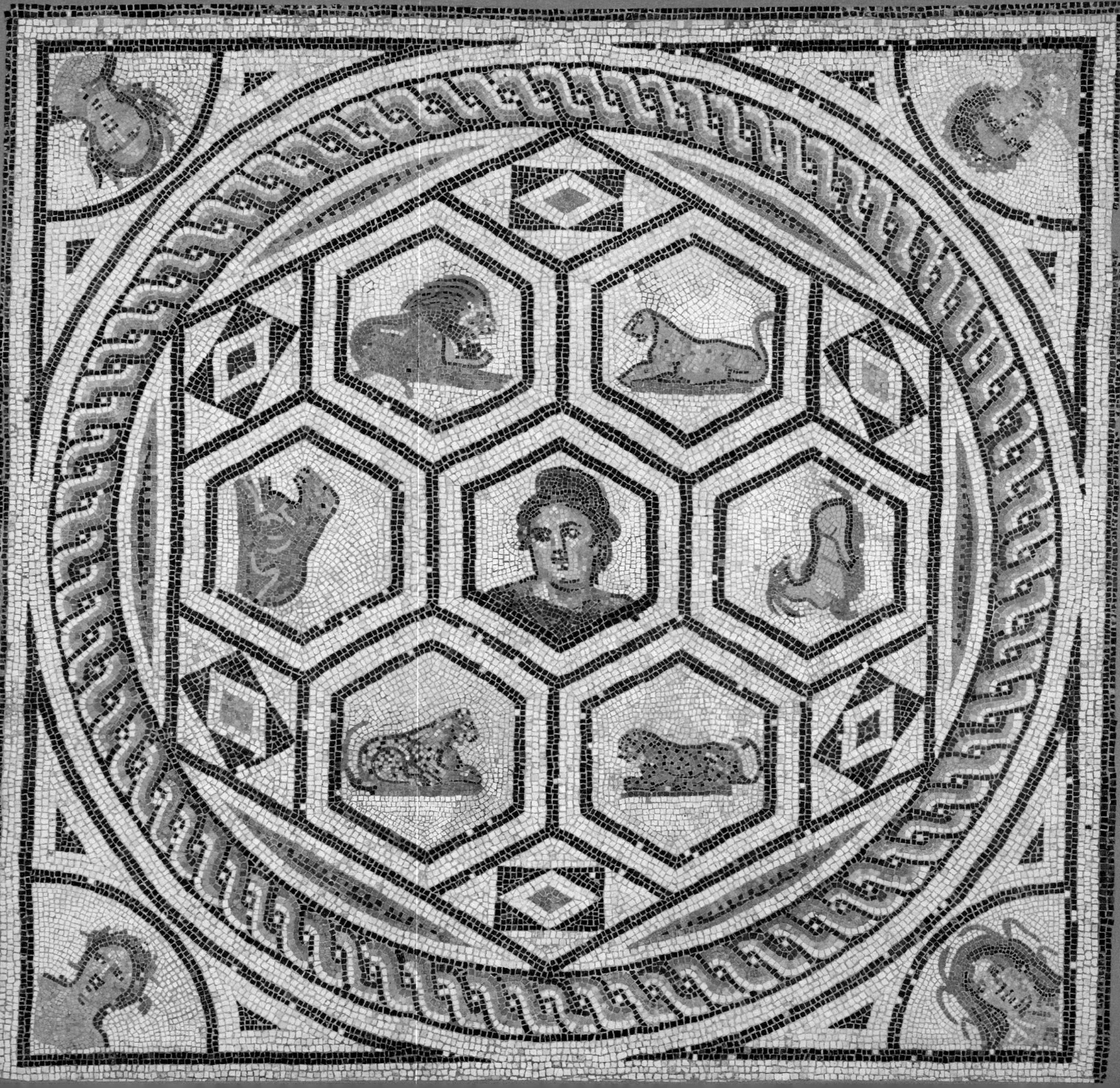

3. Mosaic Floor with Orpheus and Animals

Gallo-Roman, from Saint-Romain-en-Gal, France, AD 150–200
Stone and glass tesserae, 385.9 cm × 457 cm
62.AH.2.1–.36

Provenance

This mosaic was discovered in 1899 in a Gallo-Roman villa at Saint-Romain-en-Gal in southern France, located on the right bank of the Rhône, near Lugdunum (present-day Lyon).[1] The site was a suburb of Vienne, one of the main centers of mosaic production in the region. C. Grange, the owner of the property on which the mosaic was found, reportedly uncovered in the same area at least two more mosaics, which he unsuccessfully attempted to sell to various museums and ultimately reburied.[2] In 1911, two businessmen, Albert Vassy and Claudius Guy, became aware of the earlier excavations and purchased the property specifically to recover and sell the mosaics. In 1912, they re-excavated and lifted the Getty mosaic together with two additional pavements.[3] There is no record of the whereabouts of the Getty mosaic following its removal, but one of the other two mosaics was acquired by the British Museum in 1913, and it is likely that the Getty mosaic was offered for sale at the same time.[4] Subsequently, records of the Joseph Brummer Archive indicate that the mosaic was owned by the Brummer Gallery twice: in 1927, when it was purchased from one Widinger (or Weidinger), possibly Desire Weidinger of St. Cloud, and sold the following year to William Randolph Hearst, and in 1941, after Hearst's death.[5] J. Paul Getty purchased the mosaic in 1949 and donated it to the J. Paul Getty Museum in 1962.[6]

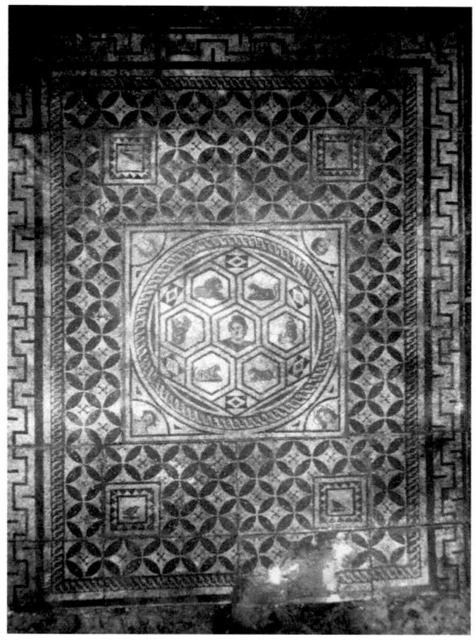

Figure 8. Excavation photo showing Mosaic Floor with Orpheus and Animals, Saint-Romain-en-Gal, France, 1912. Cliché Musées de Vienne, no. 1454

Commentary

A bust of Orpheus wearing a Phrygian cap decorates the central *emblema* (1.9 × 1.9 meters) of this mosaic.[7] The head is framed by a hexagon and bordered by six additional hexagons containing a variety of reclining animals, including a bear, a goat, a male lion and a female lion, and two other felines. This honeycomb pattern is itself enclosed by a large hexagon within a circular *guilloche* border. Geometric shapes—a diamond within a rectangle—fill the six spaces left between the hexagonal animal panels. The entire composition is bordered by a square frame with personifications of the four Seasons in the corners: Summer occupies the upper left; Fall is in the lower left; Winter is in the lower right; and Spring is in the upper right.[8] A 1912 photograph shows the floor in situ (fig. 8), including the pattern of intersecting black-and-white circles inset with crosses (also in the Getty's collection) that surrounded the *emblema*. In the corners of this geometric background are four small rectangular panels depicting birds within frames composed of triangles.[9] The outer border of the floor is decorated with a *guilloche* bordered by a meander pattern.

Orpheus, the mythical musician and poet from Thrace, was famous for his descent into the Underworld to retrieve his dead wife, Eurydice. The scene most frequently represented in Greek and Roman art, however, is the moment when every living creature gathered around Orpheus; the music from his lyre was so divine that it enchanted not only the wild beasts but even the rocks and trees of Mount Olympus.[10] The event, often depicted with Orpheus holding his lyre and surrounded by numerous animals, appears in two different iconographic traditions: narrative, with all of the figures interacting in a single space, and allusive, with Orpheus and the animals in separate compartments, usually incorporated in a larger geometric design. The Getty's mosaic exemplifies this latter, so-called multiple decor design, a local style composed of a grid-like framework filled with a series of related figures or scenes. This type of mosaic was especially popular in the upper Rhône valley in the second century AD, and the majority of examples have been found at Vienne.[11]

The Getty Museum's Orpheus mosaic was one of seven elaborate mosaic floors discovered in the same villa at Saint-Romain-en-Gal. The plan of the building, known only in a sketch preserved in the Joseph Brummer Archives, shows a series of rooms accessed by a corridor (fig. 9).[12] Three of the mosaics, including the Orpheus, were found in 1899.[13] One of these, now in the British Museum (fig. 10), is composed of several busts set within an intricate geometric design of interlacing swastikas composed of triangle patterns and guilloche bands. The central *emblema*, a modern restoration depicting Silvanus, originally represented Dionysos holding a *thyrsus* (staff topped with a pinecone) and a *krater*. Four roundels in the corners contain busts of Silenus and a maenad, Pan and a maenad, a satyr, and Dionysos.[14] The third mosaic discovered in 1899, now in the Musées de Vienne, is decorated with ornamental geometric and floral patterns arranged in hexagons and two *kantharoi* in circular medallions.[15] Investigations of the same property in 1902 uncovered four additional mosaics.[16] The most elaborate of these displays a central image of Hylas and the Nymphs (fig. 11).[17] Like the Getty mosaic, this figural scene is relatively small in comparison to the larger geometric patterns, which are surrounded by a series of half circles and spandrels decorated with shells and floral motifs. The other three mosaics were more fragmentary. One preserves a centerpiece with two birds perched on an amphora; the second is a small piece with a rectilinear pattern and a guilloche; and the third consists of a corner depicting a female head, possibly a Season, with a series of floral patterns in squares running along one edge of the mosaic.[18]

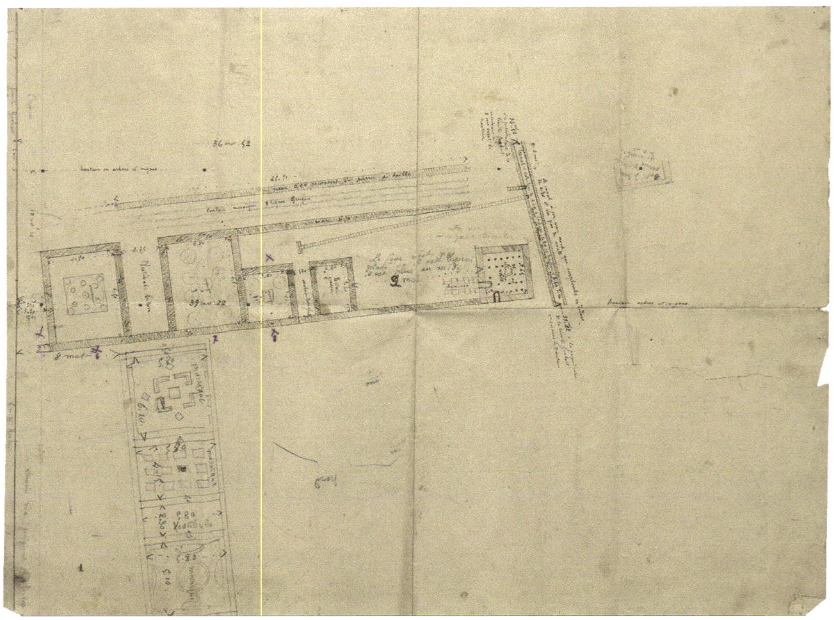

Figure 9. Sketch showing the excavation plan of the villa at Saint-Romain-en-Gal, drawn after 1902. The room with the Orpheus mosaic is on the far left. The Brummer Gallery Records, P4136

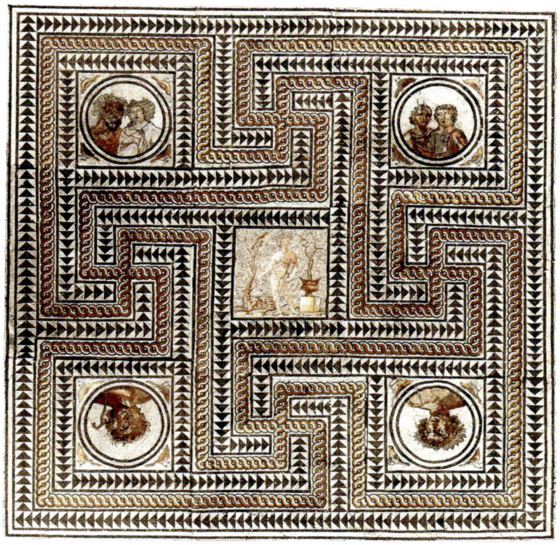

Figure 10. Mosaic Floor with Silvanus and Followers of Dionysos, Gallo-Roman, AD 150–200. Found in Saint-Romain-en-Gal, France, 1899. Stone and glass tesserae, 270 × 270 cm. London, British Museum, BM 1913,1013.1

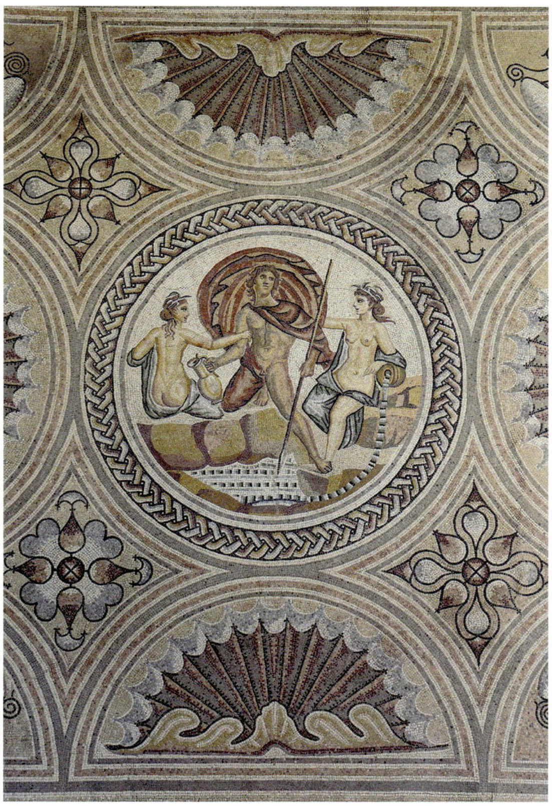

Figure 11. Detail of Mosaic Floor with Hylas and the Nymphs, Gallo-Roman, AD 150–200. Found in Saint-Romain-en-Gal, France, 1902. Stone and glass tesserae, 480 × 480 cm. Musée Gallo-Romain de Saint-Romain-en-Gal

Although some of the mythological themes represented in the mosaics of this Gallo-Roman villa are related—Orpheus was associated with the origin of the mysteries of Dionysos in Thrace, and Orpheus and Hylas joined the expedition of the Argonauts—without a complete plan or detailed description of the villa itself, it is difficult to identify an iconographic program or to understand the relationship between the mosaics and the architectural spaces they decorated.

Comparanda

A series of additional multiple decor mosaics from Vienne resembles the Getty mosaic in theme and composition.[19] One mosaic pavement also from Saint-Romain-en-Gal similarly depicts Orpheus in a square *emblema* in the center, but the surrounding honeycomb pattern with animals covers the entire floor instead of being set within a geometric background.[20] Another example composed of octagons in an arrangement comparable to the Getty mosaic was found in the House of Orpheus in the Campus Martius (Field of Mars) at Vienne, where it decorated the *frigidarium* of a private bath.[21] Indeed, the motif of Orpheus and the animals was popular on mosaics throughout the Roman Empire.[22] A variation of the type—a black-and-white mosaic divided into nine intertwined circles with Orpheus in the center—comes from Santa Marinella, near Civitavecchia, on the coast of the Tyrrhenian Sea, northwest of Rome.[23] One of the closest comparisons, however, was discovered at Thysdrus (present-day El Djem) in Tunisia.[24] Here a central medallion decorated with a bust of Orpheus in an octagonal frame is surrounded by eight additional octagons, each containing a tondo with a different reclining animal; four smaller squares between the compartments hold birds. In a similar composition from Bararus (present-day Henchir-Rougga, near El Djem), the figures occupy a series of circular wreaths arranged in a grid pattern.[25]

Condition

A photograph in the archives of the Musées de Vienne, presumably taken in 1912, shows the mosaic in situ but already cut into sections for removal.[26] At that time, it was reportedly reinforced with concrete and divided into twenty-two pieces for transport.[27] During the treatment of the mosaic that took place before it entered the collection of the J. Paul Getty Museum, many details were restored, including the animals and the shoulder of Orpheus. The Season in the upper left corner is a modern restoration based on the original, which was preserved in situ.

Bibliography

De Villefosse 1899, 102–3; Bizot 1899, 1–2; Bizot 1903, 1; Lafaye 1909, no. 219; Vassy and Guy 1913, 628–32; *Le Moniteur Viennois* 1913, 2; *La dépêche de Lyon*, 1913; *Le Progrès*, 1913, 1; *Lyon republicain*, 1913; Fabia 1923, 99; Hammer Galleries, New York 1941, 329, lot 563; Parke-Bernet Galleries, New York 1949, lot 863; Getty and Le Vane 1955, 83–84, 153, 247–48; Stern 1955, 69, no. 5; Langsner 1958, J14; Getty 1965, 20, 82–83; Stern 1971, 130–35, fig. 10; Vermeule and Neuerburg 1973, 52–53, no. 112; Lancha 1977, 72–73, fig. 29; Lancha 1981, 229–32, plates 126–27, no. 373; Armstrong 1986, 143, fig. 10; Lancha 1990, 111–12, no. 52.

A.B.

NOTES

1. The discovery of the mosaic was first published in de Villefosse 1899, 102–3, which reported the mosaic as having been found at the nearby site of Sainte-Colombe. Early newspaper articles, however, describe Saint-Romain-en-Gal as an enclave of Sainte-Colombe, possibly accounting for the confusion. For more on this identification, see Stern 1971, 131–35.
2. Grange's farmer, M. Prost, discovered the mosaic while working in a vineyard on the property.
3. Lafaye 1909, no. 219. Vassy and Guy 1913, 628–32, indicate that the mosaic was uncovered and removed in 1912.
4. The British Museum purchased one of the mosaics in 1913 (inv. no. 1913, 1013.1).
5. The Metropolitan Museum of Art, New York, has digitized the Brummer Gallery Records, now available online on the museum's Thomas J. Watson Library website; see Digital Collections from the Metropolitan Museum of Art, Brummer Gallery Records, http://libmma.contentdm.oclc.org/cdm/landingpage/collection/p16028coll9. Catalog cards recording purchase and sale information show that the mosaic was owned once under the number P4136 and once as N5207. The Hearst estate sold the mosaic in 1941 at the Hammer Galleries, New York, lot 563.
6. The Brummer estate sold it in 1949 at Parke-Bernet Galleries, New York, lot 863.
7. Fragment no. 62.AH.2.13.
8. According to Lancha, who dates the mosaic to AD 150–200; see Lancha 1981, 229–32, no. 373.
9. Lancha identifies the birds as a partridge, a duck, a pheasant, and a magpie; see Lancha 1981, 230.
10. For an in-depth examination of the image of Orpheus in Roman mosaics, see Jesnick 1997.
11. For a general discussion of the multiple decor style, see Dunbabin 1999, 74–76, with additional references.
12. Although the original catalog card for Brummer Archive number P4136 does not appear in the online collection of the Metropolitan Museum of Art, supplemental material linked to the record includes a hand-drawn plan of the excavations that indicates the locations of the mosaics. Henri Stern, in his summary of the history of the mosaics, records that the original architectural plan, now lost, was made by the architect M. Rambaud; see Stern 1971, 124.
13. Vassy and Guy 1913, 628–32; and Lafaye 1909, nos. 219, 220, 221.
14. Lafaye 1909, no. 220; Vassy and Guy 1913, 631; Stern 1971, fig. 1; and Lancha 1981, 232–36, plates 128–30 (images of the mosaic before restoration), no. 374, which dates the mosaic to AD 150–200.
15. Lafaye 1909, no. 221; Vassy and Guy 1913, 629–31; Stern 1971, fig. 12; and Lancha 1981, 236–39, plate 131, no. 375, dated to AD 150–175.
16. Lafaye 1909, nos. 222,1 and 2, 223, 224; Vassy and Guy 1913, 628–32. According to Vassy and Guy, a poorly preserved mosaic decorating the floor of the corridor, featuring a "simple Greek" design, was left in situ at this time.
17. Lafaye 1909, no. 224; Vassy and Guy 1918, 628; and Lancha 1981, 241–44, plates 132–33, no. 380, dated to AD 175–200, now in the Musée Gallo-Romain de Saint-Romain-en-Gal. Hylas, who sailed with Herakles and the Argonauts, was carried off by nymphs when drawing water from a spring (as in the version by Apollonius Rhodius, *Argonautica* 1.1207–1239). Although the subject of Hylas and the Nymphs was frequently depicted in Roman art, this is the only known example from Gaul.
18. Birds on amphora: Lafaye 1909, no. 222,1; and Lancha 1981, 244–45, plate 124a, no. 381 (Musée de Grenoble). Rectilinear pattern: Lafaye 1909, no. 222,2; and Lancha 1981, 245, plate 124b, no. 382 (Musée de Grenoble). Female head: Lafaye 1909, no. 223; and Lancha 1981, 246–48, plate 135, no. 383 (Musées de Vienne).
19. For a similar composition, see Lafaye 1909, no. 217. Found in 1894 in Sainte-Colombe, this design also consists of a square *emblema* with a honeycomb pattern set within a geometric background, but an image of Venus is in the center. This mosaic was originally identified as the Orpheus mosaic in the Getty Museum; the report by M. Stothart on this misidentification is published in Stern 1971, 131–33.
20. Lafaye 1909, no. 201 (repeated as no. 242); Stern 1955, 68–69, no. 4; Stern 1971, 138, fig. 15, a reconstruction of the mosaic, which was discovered in 1822 in a vineyard in Montant and then destroyed in a fire in 1968; and Lancha 1981, 226–29, plates 124–25, no. 372, which dates the mosaic to the early third century AD.
21. Lafaye 1909, no. 181 (repeated as no. 233); Stern 1955, 69, no. 6; and Lancha 1981, 89–93, plates 34–37, no. 282, which dates the mosaic to the end of the second century AD. The mosaic was found in 1859 in the southeast corner of the Campus Martius.
22. See Stern 1955, 68–77, for a catalogue of forty-seven mosaic floors from all areas of the Roman Empire with the subject of Orpheus and the animals. (The Getty mosaic is no. 5.)
23. Stern 1955, 70, no. 15, with additional references, dates the mosaic to no later than the second century AD. Stern includes two other variants on the type, with Orpheus completely separated from the animals; see Stern 1955, 73–74, no. 28 (La Chebba, Tunisia) and no. 32 (Tanger, Algeria).
24. Foucher 1960a, 8–11, plates 1, 2, end of the second century AD, from the Bir Zid district.
25. Slim 1987, 210–11, no. 78 (now in the El Djem Archaeological Museum), late second or early third century AD.
26. See Stern 1971, fig. 10 (photograph no. 1454 in the archives of the Musées de Vienne), probably taken in 1912, when the mosaic was uncovered for the second time.
27. Vassy and Guy 1913, 631, specify that of the three mosaics removed, the Orpheus was reinforced with cement to preserve it for future examination.

Villelaure

History of the Excavations

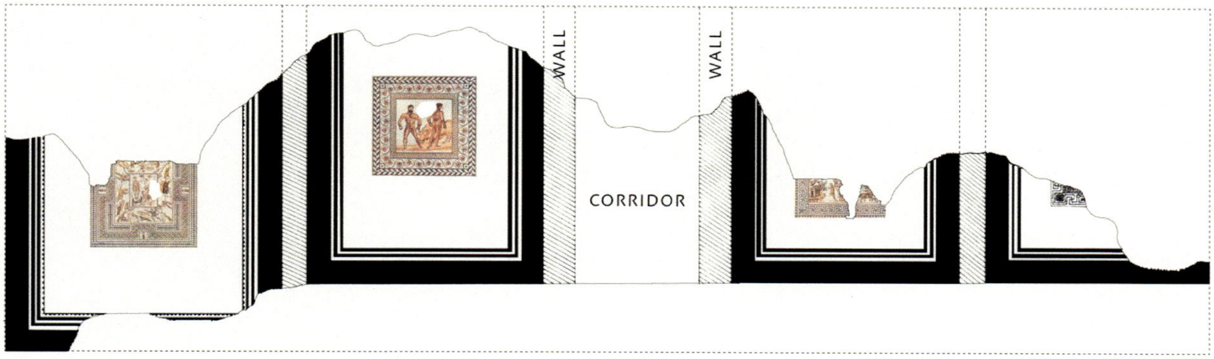

Figure 12. Floor plan of the Gallo-Roman Villa at Villelaure. Drawing by R. Prudhomme, based on an original excavation drawing by the archaeologist L.-H. Labande (see Lavagne 1977, fig. 1) with the addition of watercolors by the architect Henri Nodet (see Lafaye 1909, nos. 103–5). Composite image created by Michael Lira

The modern history of the Gallo-Roman villa located near the town of Villelaure, twenty miles north of Aix-en-Provence (the Roman town of Aquae Sextiae), begins with its initial discovery in 1832 and spans more than 180 years. The villa, located in a district known as the Tuilière, which is bordered to the southeast by the Marderic River, was covered by an uncultivated field until the land was developed for sugar beet production in the 1830s.[1] Four rooms with intact mosaic floors were uncovered in 1832, but they were reburied and not fully documented until their accidental rediscovery in 1898 and further investigation in 1900.[2] The excavation unearthed four mosaics that paved a suite of adjacent rooms identified as the residential section of a wealthy private villa (fig. 12). According to the plan made by L.-H. Labande, the four adjacent rooms were oriented north–south (measuring a total length of 23.5 meters, with a width of 7 to 8 meters) and separated by a corridor 2.5 meters long with two rooms on either side.[3] The mosaics of all four rooms had the same basic composition: a large central panel with a geometric border set on a plain white field, and, along the walls, a black border with alternating black-and-white lines. In the southernmost room (3.3 × 3.1 meters), only the badly damaged northwest corner of the pavement was preserved. The fragmentary remains included the black outer border and a corner of the geometric design that originally framed the central panel. This border was composed of a polychrome *guilloche* surrounded by a geometric design with small panels, possibly a meander pattern of swastikas and squares; a marine animal (possibly a fish or a dolphin) in one of the panels; and a rosette decorating the corner.[4] The mosaic in the second room (7 × 4.6 meters), was also fragmentary, with only the lower portion of the central panel preserved, and it illustrated a Nilotic landscape (fig. 13).[5] The preserved fragments depict, on the left, a temple perched on a cliff, a palm tree with dates, and an ibis. A crocodile is shown in the water, and, to its right, a pygmy stands on the bow of a boat, bearing an oval shield and brandishing a weapon at the crocodile. A second pygmy, also carrying a shield, stands in the center foreground. A hippopotamus appears on the far right, and, in the background, there is a cliff with the base of a building. An earlier description records images of horsemen and elephants that are no longer preserved.[6] Labande reported that there was a fragment of marble revetment in situ in the southwest corner of the room. The corridor was unpaved at the time of the 1900 excavation.[7] The Getty mosaic of Dares and Entellus (cat. 4) decorated the floor of the room (7 × 4.7 meters) on the opposite side of the corridor, and the north room (7 × 5.5 meters) contained the Diana and Callisto mosaic now in the Los Angeles County Museum of Art (fig. 14).[8] The central panel of Diana and Callisto is surrounded by a rectangular frieze depicting different hunting scenes on each side and trees in the four corners: below, two hunters trap a hare in a net; on the sides, hunters attack deer and a wild boar; and above, a man with two spears fights beasts of the amphitheater.[9] While all of the mosaics are dated to the late second century AD, the villa itself was occupied between the second and fifth centuries AD.

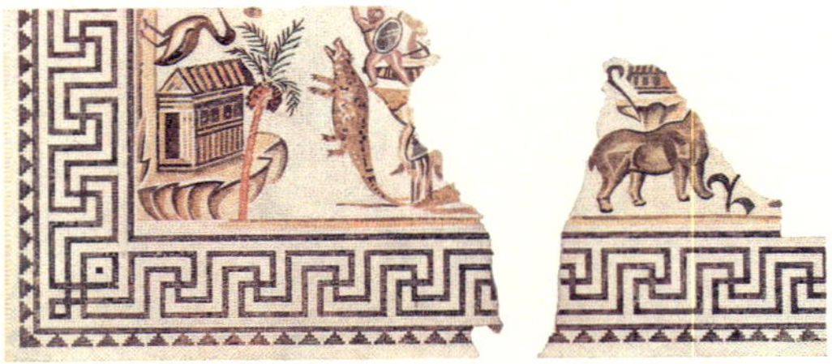

Figure 13. Henri Nodet (French, 1855–1940), watercolor (now lost) of the mosaic floor with a Nilotic landscape, 1903. Mosaic, Gallo-Roman, AD 175–200. Found in Villelaure, France, 1832

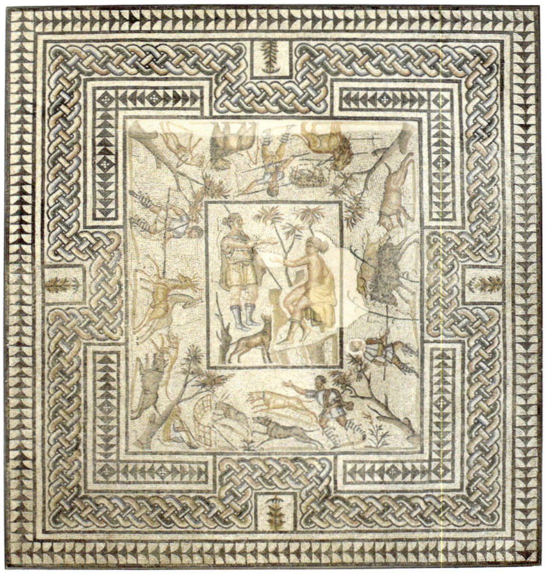

Figure 14. Mosaic Floor with Diana and Callisto Surrounded by Hunt Scenes, Gallo-Roman, AD 175–200. Found in Villelaure, France, 1832. Stone and glass tesserae, 300 × 270 cm. Los Angeles County Museum of Art, M.71.73.99

This exceptional site recently became the focus of a new project of investigation and preservation owing to the successful efforts of the community of Villelaure, located just southeast of the ancient villa.[10] When the site was threatened by modern construction in 2006, the local villages, including Villelaure, Lauris, and Ansouis, intervened to preserve the area. The association of Villa Laurus en Luberon organized a diagnostic survey to assess the extent of the archaeological site. Subsequent excavations confirmed the existence of an extensive Gallo-Roman villa, and a proposal was made for its preservation in view of its great archaeological and historical value. Under the auspices of the Institut National de Recherches Archéologiques Préventives (Inrap), Robert Gaday unearthed additional foundations of the villa, identifying both residential and agricultural areas (fig. 15).[11]

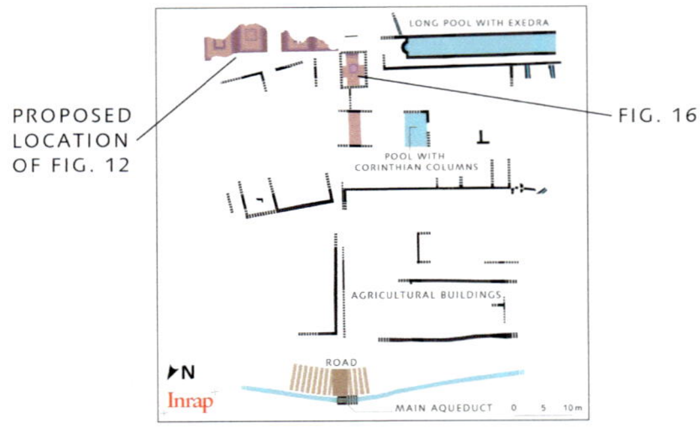

Figure 15. Plan of the 2006 excavations of the Gallo-Roman Villa at Villelaure with proposed locations of the mosaics excavated in the nineteenth century. Plan provided by Robert Gaday and Frédéric Gueriel (Institut National de Recherches Archéologiques Preventives)

This type of villa, known as a *villa rustica*, or countryside villa, was often the center of a large agricultural estate composed of separate buildings to accommodate laborers, animals, and crops. At Villelaure, amenities of the main building included a peristyle court and a long pool with an exedra. The survey of the surrounding area revealed the existence of a water channel with an underground aqueduct, a ceramic kiln, and the remains of a necropolis. The excavations also uncovered a fifth mosaic decorating the floor of a rectangular room (5.7 × 3.8 meters). This fragmentary mosaic (fig. 16) preserves the lower part of a figure wearing shoes and a cloak, which was framed by a circular guilloche inscribed within a square. Notably, the recently discovered mosaic has the same format as the other four: a framed central figural scene (1.7 meters square) surrounded by a white field with a black-and-white border along the walls.

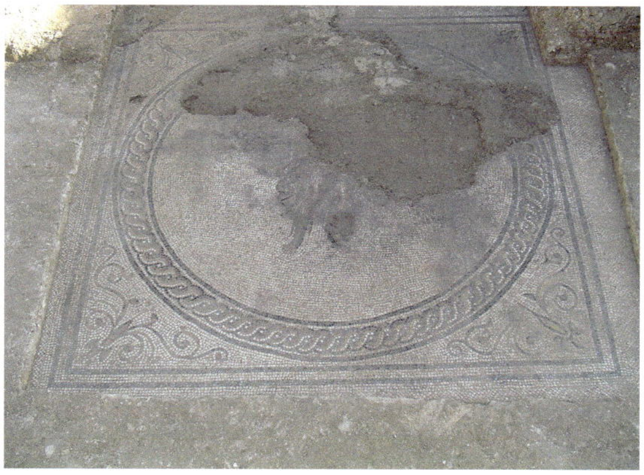

Figure 16. Excavation photo showing the central panel of the mosaic with a male figure from the Gallo-Roman villa, Villelaure, France, 2006. Provided by Robert Gaday (Institut National de Recherches Archéologiques Preventives)

The 2006 excavations did not confirm the original location of the four previously excavated mosaics.[12] At present, the focus remains on preserving the site from further modern damage, with the intention of carrying out additional excavations in the future.

A.B. and N.B.

NOTES

1. For additional details on the discovery and later history of the mosaics at Villelaure, see cat. 4, in the present volume.
2. For a discussion of the 1832 discovery, see Jacquème 1922, 101. The excavation report is published in Labande 1903, 3–13; and de Villefosse 1903, 13–14, 20–23. See also Lavagne 1977, 177–78; Lavagne 2000, nos. 913–16; and the detailed summary with bibliography in Tallah 2004, 374–75.
3. A drawing of the floor plan is included in Labande 1903, 5, and reproduced in more detail in Lavagne 1977, 178, fig. 1, and Lavagne 2000, 307, fig. 48.
4. The description is based on the report of L.-H. Labande; see Labande 1903, 6–7. According to Lafaye 1909, no. 102, the fragments of this mosaic were donated to the Académie de Vaucluse for the collection of the Musée Calvet d'Avignon, but they are now lost. See also Lavagne 2000, no. 913, plate C.
5. Although this mosaic is now missing, it is known from watercolors made by the architect Henri Nodet prior to its disappearance and published in Lafaye 1909, no. 103; Labande 1903, plate 2; and Lavagne 2000, no. 914, plate C.
6. *Notice sur la terre de Villelaure et Janson exposée en vente à l'enchère publique devant le Tribunal de première instance séant à Apt.*, 1848, 15, cited in Lavagne 2000, 309.
7. The 1903 report states that the pavement had "entirely disappeared" at the time of excavation and that there were no recorded documents of a paved floor; see Labande 1903, 7.
8. Lafaye 1909, nos. 104 (Dares and Entellus) and 105 (Diana and Callisto). The Diana and Callisto mosaic was displayed together with the Dares and Entellus mosaic at the exhibition *Roman Mosaics across the Empire*, held at the Getty Villa from March 30, 2016, to September 12, 2016.
9. The wild beasts in the top scene possibly represent a lion, a bear, a panther, and a lioness; see Lavagne 2000, 313–15.
10. Many thanks to our colleague André Girod and to the association of Villa Laurus en Luberon for all their assistance in gathering information about the history, both ancient and modern, of the site of Villelaure. In conjunction with these investigations, the association initiated a project to transform the nearby eighteenth-century Château-Verdet-Kleber in Villelaure into a museum and workshop for the study of mosaics.
11. Gaday et al. 2006.
12. According to the 2006 report, the north–south alignment of the four mosaics documented by Labande in 1903 does not correspond with the orientation of the fifth mosaic; see Gaday et al. 2006, 38.

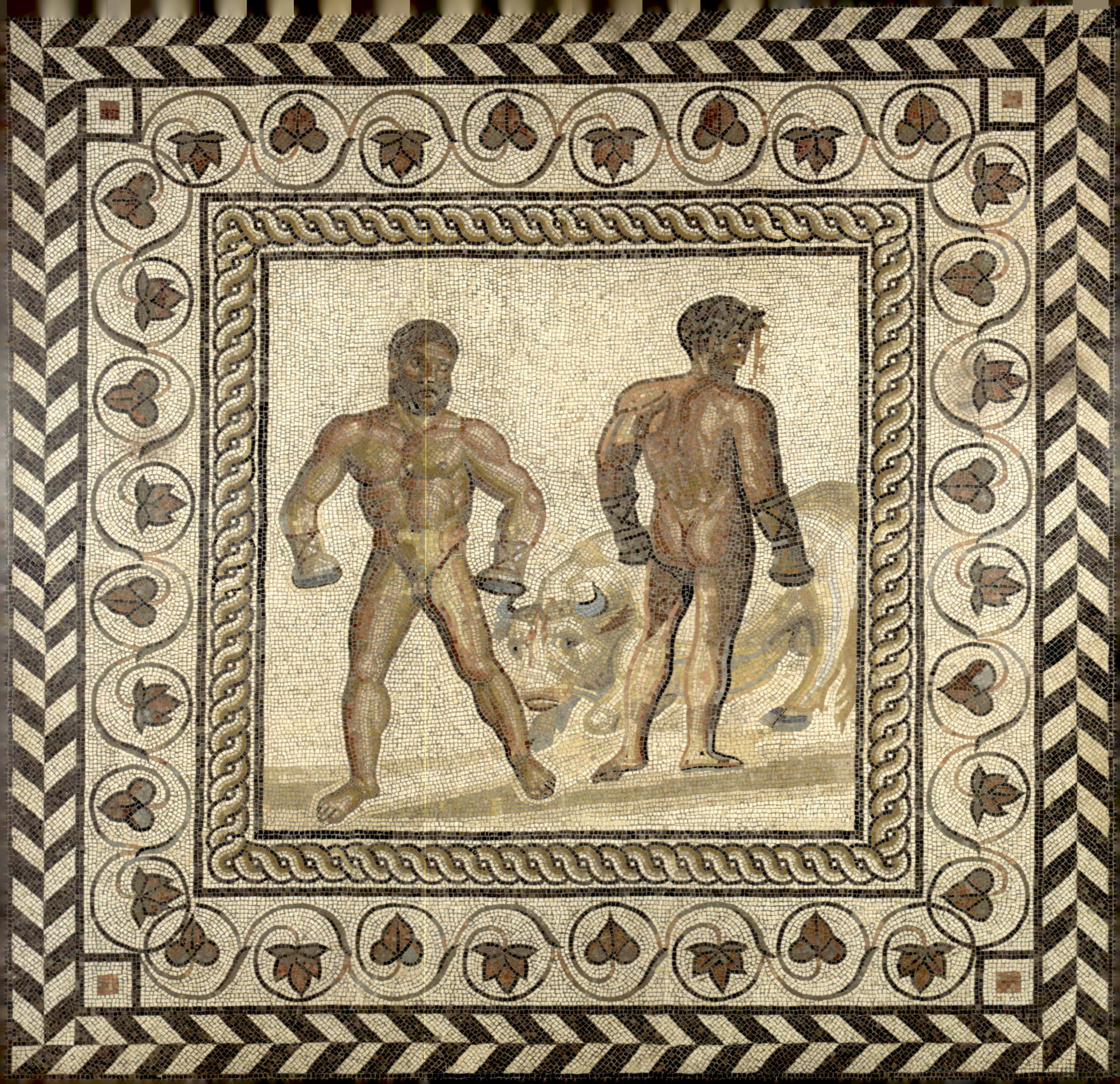

4. Mosaic Floor with Combat between Dares and Entellus

Gallo-Roman, from Villelaure, France, AD 175–200
Stone and glass tesserae, 208 cm × 208 cm
71.AH.106

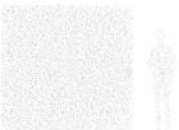

Provenance

This mosaic was found in a Roman villa near the modern town of Villelaure.[1] In 1836, M. Aliqué, the steward of the Marquis de Forbin-Janson, unearthed four rooms with mosaic floors divided by a corridor. The mosaics were soon reburied and their exact location forgotten.[2] In the winter of 1898, the villa was accidentally rediscovered by Pierre Raynaud, whose plow brought up pieces of concrete with mosaic decoration.[3] The landowner, M. Peyrusse, organized excavations in the spring of 1900 and uncovered the four rooms, their mosaic floors in various states of preservation. The floor of the first room was heavily damaged; the second room preserved a small section of a Nilotic scene (see fig. 13); the third contained the Getty mosaic of Dares and Entellus; and the fourth had a partially preserved mosaic of Diana and Callisto surrounded by a hunting frieze (see fig. 14).[4] Visible to the public in situ after their discovery, the central panels of the mosaics were lifted by 1913 and stored at the farm in Roules owned by M. Peyrusse.[5] By 1920, they were sold to Frédéric Rinck, a dealer in Avignon, and then bought by R. Ancel, a dealer in Paris.[6] The Louvre attempted to acquire the mosaics in 1923, but they were ultimately sold to the New York art dealer Joseph Brummer in 1926.[7] Purchased soon after by William Randolph Hearst, the mosaics were transferred to Hearst Castle, San Simeon, in May 1926. Upon Hearst's death, they were bought by his estate manager, James N. Evans, who sold the mosaic of Dares and Entellus to J. Paul Getty in 1971.[8]

Commentary

This square mosaic panel depicts two nude boxers standing in front of a large white bull. The central image is surrounded by a three-part border: an inner *guilloche* band, a rinceau of alternating vine and ivy leaves with red squares in the four corners, and an outer border of black-and-white chevrons.[9] The panel decorated the center of a large white mosaic floor with a border of white and black lines (4.7 by 7 meters), which is now lost but documented in excavation drawings. The mosaic is composed of tesserae made primarily of stones in shades of red, yellow, and black, except for the horns of the bull, which are made of pale blue glass.

The subject of the mosaic is the boxing match between Dares and Entellus, as described in Virgil's *Aeneid* (5.362–484).[10] In Virgil's account, the event takes place during the funeral games held by Aeneas, the future founder of Rome, to honor his late father, Anchises. Dares, a young Trojan in Aeneas's company, famous for his strength, offers to compete and calls for an opponent. Entellus, an older local Sicilian, takes up the challenge and, despite his age, soundly defeats Dares, nearly beating him to death before Aeneas intervenes to end the fight. Entellus is named the victor and brazenly kills his prize, a large white bull, with a single blow to demonstrate his strength and honor the gods. The Getty mosaic illustrates the conclusion of the episode: Entellus, the bearded, stocky figure on the left, has just struck the bull and stands with his chest thrust forward and his arms flexed. He glances back toward Dares, who turns away, his arms lowered and his head bleeding. Both men wear *caesti*, strips of leather weighted with lead or iron, wrapped around their hands and forearms. These weighted gloves were distinctively Roman, and Virgil describes them in detail.[11] The bull is shown at the moment of death: its forehead bloody and its forelegs crumpled under the force of Entellus's blow.

The Getty mosaic originally decorated the floor of a long room, perhaps a *triclinium*, with its entrance likely on the eastern end facing the Marderic River.[12] As is usual, the orientation of the mosaic favors the view of the homeowner and his guests from within the room. Upon entering the space, the Dares and Entellus mosaic would appear upside down, and one would have had to walk over or around the mosaic to the western end of the room in order to view it properly.[13] The iconography, especially when considered with that of the adjacent rooms containing the Diana and Callisto and the Nile mosaics—which are the only known examples of their kind in Gaul—likely served to emphasize the patron's awareness of Roman literature and culture.[14] Although the identity of the patron is unknown, he was probably a wealthy member of the elite with connections to Aquae Sextiae (present-day Aix-en-Provence), a center of mosaic workshops in southern Gaul.[15]

Given the unique combination of mosaic pavements found in this villa at Villelaure, it is likely that the iconography of these mosaics had personal or local significance. The Getty mosaic draws on the repertoire of figural mosaics from this region, which often feature scenes from literature and myth.[16] The mosaics from Villelaure, however, also resonate with activities of contemporary life in southern Gaul. For example, the mosaic showing Diana and Callisto (see fig. 14) illustrates a scene from Ovid's *Metamorphoses* featuring Diana, patron goddess of the hunt, in the central panel, while the surrounding border contains a frieze of various hunting scenes. Hunting was a notable elite pastime and a spectacle in the local amphitheaters.[17] In a similar fashion, the mosaics with Dares and Entellus may have reminded viewers of contemporary athletic games as well as the passage in the *Aeneid*.[18] Collectively, the mosaics speak to an interest in Roman literature, local leisure activities, and entertainment—appropriate themes for a countryside villa.

Comparanda

Despite the widespread popularity of Virgil's works, representations of the *Aeneid* are surprisingly uncommon in mosaics.[19] Furthermore, the boxing match between Dares and Entellus is a minor episode in the *Aeneid* and a seemingly odd choice for a mosaic floor.[20] This particular scene, however, proved popular in southern Gaul, where it occurs on at least four other mosaics—three from Aix-en-Provence and one from Nîmes, all dating to the late second century AD.[21] All five mosaics share the same basic composition—Entellus and Dares flanking the dying bull—but they vary in quality and in the treatment of the surrounding design elements.[22]

Condition

The mosaic is in good condition with few losses except for a small section above the left shoulder of Dares, which was missing upon its discovery and restored sometime after the mosaic was lifted.[23] The surface of the mosaic was ground down and polished, likely around the same time as the other restorations.

Bibliography

Bulletin de la Société nationale des antiquaires de France 1900, 167; Comptes-rendus de l'Académie des inscriptions et belles-lettres 1900, 222; Labande and de Villefosse 1900, 113, no. 3; Gauckler 1901, 340; Lafaye 1901, 117–18, 119; Fowler 1902, 370; Labande 1903, 8–10, 5; de Villefosse 1903, 20–23, no. D, plate 2; Lafaye 1909, 24, no. 104; Jacquème 1922, 101–6, plate 8; Reinach 1922, 278, no. 5; Michon 1923a, 109; Michon 1923b; Michon 1923c, LX; Sautel 1939, 4, no. 11; Vermeule and Neuerburg 1973, 54–55, no. 114; Fredericksen 1975, 55; Lavagne 1977, 177–82, fig. 2; Lavagne 1978, 135–46, fig. 2; Rey 1979, 17–21; Ministère de la culture et communication 1981, 349, fig. 367; Geymonat 1987, 156, fig. 16; Seignoret 1987, 4; Egea 1988; Darmon 1990, 70, 73–74; Lavagne 1994, 213–14, fig. 8; Darmon 1995, 62–63; Lancha 1997, 111–13, no. 57, plate 61; Lavagne 2000, 306–8, 310–11, fig. 48, no. 915, plate 101; Tallah 2004, 374–75, fig. 469; Gaday et al. 2006, 18–19, fig. 4; Sauze and Muret 2008, 11–18; Lavagne 2014, 197, fig. 272.

N.B.

NOTES

1. The site was in a field known as the Tuilière, two kilometers north of Villelaure, twenty-five meters west of the Marderic River, under less than a meter of earth. The excavation report is published in Labande 1903, 3–13, and de Villefosse 1903, 13–14, 20–23. See also Lavagne 1977, 177–78. Recent diagnostic excavations at the site have unearthed new finds; see Gaday et al. 2006, 30–31, 38.
2. According to local traditions, pieces of mosaic floors with representations of humans and animals were found in the mid-nineteenth century, but they are now lost; see Labande 1903, 3. By 1872, locals believed that the mosaics had been destroyed due to the plowing of the field; see Jacquème 1922, 101.
3. See Labande 1903; on Pierre Raynaud, see Seignoret 1987.
4. Now in the Los Angeles County Museum of Art (M.71.73.99). For a discussion of the other mosaics and excavations at the site, see "Villelaure: History of the Excavations," in the present catalogue.
5. The mosaics may have been lifted and set in cement as early as 1901 (see Gauckler 1901, 340) and only moved off-site by 1913; see Labande 1903, 4n1; Rey 1979, 25; and Lavagne 2000, 306. According to Labande, the mosaic fragments from the first room were donated to the Academie de Vaucluse and intended for the Musée Calvet d'Avignon, but according to Lavagne, there is no record of the donation in the archives; see Lafaye 1909, no. 102; Rey 1979, 25; and Lavagne 2000, 306. The pavements that surrounded the central scene are now lost; see Labande 1903, 4n1.
6. See Rey 1979, 25; and Lavagne 2000, 306.
7. In Lavagne 1977, 178n10, the author cites a letter from E. Michon (Michon 1923b), antiquities curator at the Louvre at the time, concerning the mosaics. The fragmentary Nilotic mosaic may have been sold to Brummer and damaged in transit to the United States (Lavagne 2000, 307); however, it may have been lost earlier because Étienne Michon notes that only two mosaics were with the dealer in Paris in 1923 (Michon 1923a, 109). These mosaics are recorded in the Brummer Gallery Records (N1101: Large Gallo-Roman mosaic and N1102: Large Gallo-Roman mosaic). The Metropolitan Museum of Art, New York, has digitized the Brummer Gallery Records, now available online on the museum's Thomas J. Watson Library website; see Digital Collections from the Metropolitan Museum of Art, Brummer Gallery Records, http://libmma.contentdm.oclc.org/cdm/landingpage/collection/p16028coll9.
8. The Diana and Callisto mosaic was sold to Phil Berg in 1962, and upon his death in 1983 it was donated to the Los Angeles County Museum of Art; see Lavagne 2000, 307.
9. See Lavagne 2000, 216. This type of chevron design is relatively rare; comparanda are found in Aix-en-Provence (Lavagne 2000, no. 742), Orange (Lavagne 1979, nos. 45, 58), Nîmes (Esperandieu 1935, nos. 6, 11, 35), and Russi, Italy (Lancha 1977, fig. 78).
10. The composition departs slightly from Virgil's description in order to condense several moments into one recognizable scene. In the poem, after Aeneas stops the fight (*Aeneid*, 5.461–64), Dares is carried away by his companions (*Aeneid*, 5.468–70); Entellus is awarded and kills the prize bull only after Dares departs (*Aeneid*, 5.474–76). Virgil's account of the boxing match imitates Homeric and Hellenistic models (see Homer's *Iliad*, 23.651–99, and *Odyssey*, 18.1–108; Apollonius's *Argonautica*, 2.30–97; and Theocritus's *Idyll*, 22).
11. Virgil's *Aeneid*, 5.405–13. On *caesti*, see Poliakoff 1987, 68–79.
12. This proposed orientation is based on Labande's 1903 plan (see fig. 12) and on comparable mosaic floors, including two with the Dares and Entellus mosaics from Aix-en-Provence (rue des Chartreux and rue des Magnans), which favor a view of the mosaic from within the room; see Lavagne 2000, nos. 840, 857, fig. 45.
13. The black-and-white chevron border, which encircles the mosaic in a counterclockwise direction, would perhaps have encouraged movement around the mosaic; see Clarke 1979, 21, on "kinesthetic address" in Roman mosaics.
14. The Diana and Callisto mosaic illustrates a specific and rarely portrayed episode of the myth of Diana's discovery of the pregnant Callisto (Ovid, *Metamorphoses*, 2.401–530). Callisto is exiled, transformed into a bear, and nearly killed by her own son during a hunt, but is ultimately rescued and transformed into a constellation. The wounded bear depicted in the hunting frieze above the central panel may have reminded viewers of Callisto's imminent transformation; see Lavagne 2000, 309, 313, nos. 914, 916.
15. See Lavagne 2000, 291, 310. The owner of the villa in Villelaure may have seen representations of Dares and Entellus on mosaics in Aquae Sextiae and requested the motif; three of these mosaics are dated slightly earlier, AD 130–150.
16. Examples from southeast Gallia Narbonensis include Theseus and the Minotaur (Lavagne 2000, no. 788), Orpheus (Lavagne 2000, nos. 511, 764), Bacchus and Ariadne (Lavagne 2000, no. 835), Narcissus (Lavagne 2000, no. 629), the labors of Hercules (Lavagne 2000, no. 554), and a theater scene (Lavagne 2000, no. 787, perhaps from Terence's *Adelphi* or a play by Menander).
17. Lavagne 2000, 311–15, no. 916, plates 102, 103. Ovid, *Metamorphoses*, 2.401–530. The hunting scene includes animals that were actually hunted in southern Gaul (the hare and the deer) with animals more typically found in *venationes* of the amphitheater (the bear, the lion, and the leopard); compare mosaics from Lillebonne (Darmon 1994, no. 885), Vallon (Rebetez 1992, 15–29), and Nennig (Dunbabin 1999, 82, fig. 84).
18. Athletic spectacles, which included boxing matches, were likely held in arenas, amphitheaters, and *palaestrae*. In his *Epistulae* (Letters), Pliny the Younger notes the popularity of *gymnicus agon* in Gaul, as far north as Vienne (*Epistulae*, 4.22); for a review of athletics in Gaul, see Newby 2005, chapter 3, and König 2005, chapter 5, especially 219–22.
19. Other allusions to the *Aeneid* found in mosaics of Roman villas, include the Dido and Aeneas from Low Ham, Somerset; Aeneas with the Golden Bough from Frampton, Dorset; Europa and the bull from Lullingstone, Kent; and representations of Virgil from Sousse (Dunbabin 1978, 131, 242, plate 130) and Trier (Landesmuseum, inv. 10703-24; Parlasca 1959, 41). See Dunbabin 1999, 96–98, for references on the mosaics from Britain, and Ling 1998, 73 and 75, for a discussion of Roman legend in mosaics.
20. The boxing match would have been recognizable to an ancient viewer (compare a graffito from Pompeii that quotes *Aeneid*, 5.389, honoring Entellus *CIL* IV 8379). See Milnor 2014, 269, no. 44; Ferraro 1982, 31, no. 22; and Lavagne 2000, 290.
21. The Aix-en-Provence examples include a mosaic discovered near the Hôpital Saint-Jacques in 1790, recorded in a watercolor but now lost (Lavagne 2000, no. 789, plate 81), a second from rue des Chartreux in 1988 (Lavagne 2000, no. 840, plates 90, 91), and a third from rue des Magnans in 1992 (Lavagne 2000, no. 857, plates 93–98). The Nîmes mosaic differs in its composition since it includes Aeneas presiding as judge; see Lancha 1997, nos. 50, 101, and Darmon 1990, 73. For discussions of the five mosaics, see Lavagne 1994, 211–15; and Lavagne 2014, 196–98. A possible sixth mosaic, in Arles, is damaged and currently unpublished; see Lavagne 1990, 22. The only other known representations of the scene are two funerary reliefs from northern Gaul (Lefebvre 1987, 14–17; and Espérandieu 1913, no. 4339, reproduced in Lavagne 1994, figs. 9 and 10) and a heavily restored relief in the Lateran collection (Helbig 1963, 731, no. 1016).
22. By the second century AD, Gallo-Roman mosaic workshops were well established and had developed local styles, such as the geometric grid design known as "multiple decor" that was especially popular in the upper Rhône valley; see Dunbabin 1999, 74–76. A high-quality example of this pattern is the Dares and Entellus

mosaic from rue des Magnans in Aix-en-Provence; see Lavagne 2000, no. 857, plate 94.

23. The missing section is visible in watercolors by Henri Nodet (see de Villefosse 1903, plate 2, 32; and Lafaye 1909, plate 104) and also documented in a photograph by Franki Moulin (see Lafaye 1901, 117, 119). Restorations may have been made in the United States during the 1930s; see Lavagne 2000, 310.

NORTH AFRICA

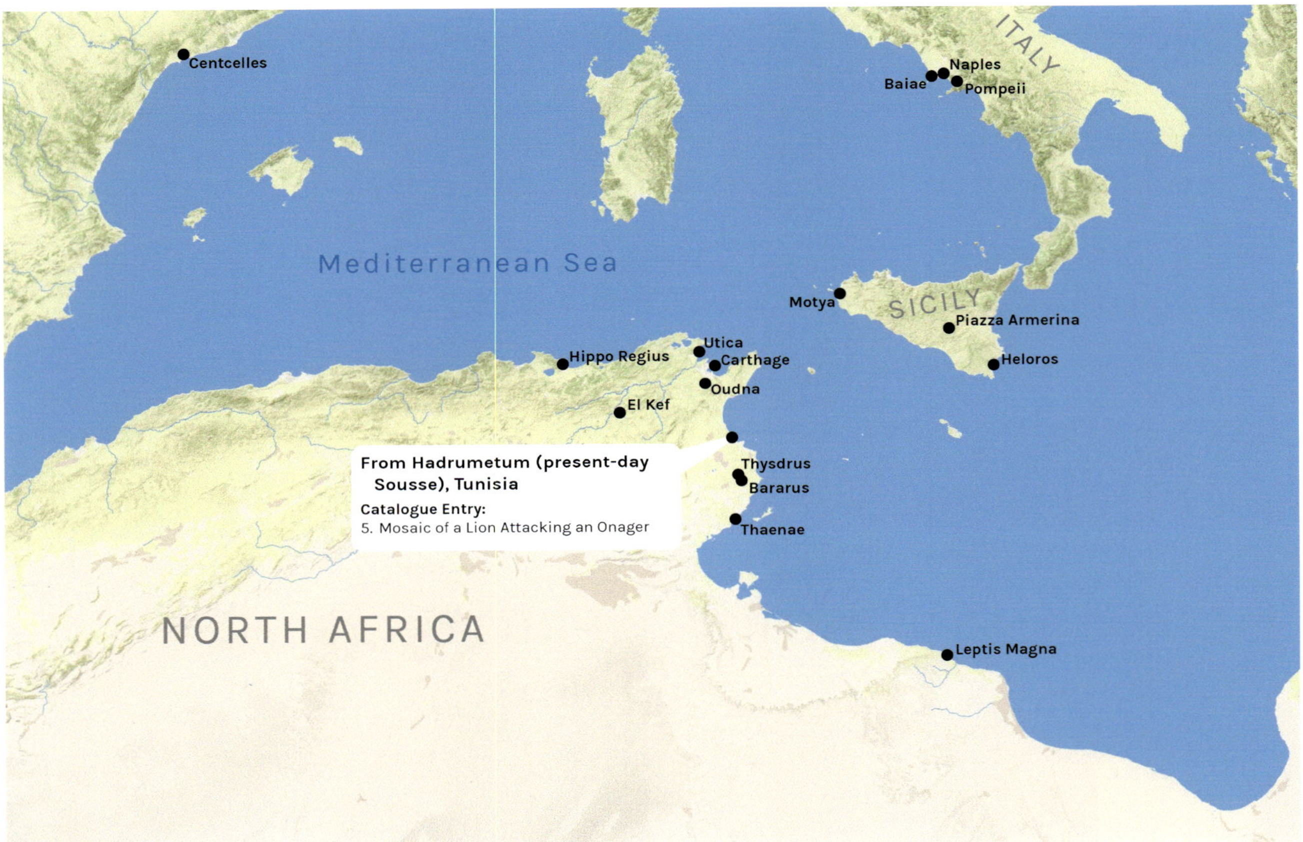

North Africa

More mosaics have been preserved in the Roman provinces of North Africa than anywhere else in the empire, especially in the prosperous agricultural province of Africa Proconsularis (present-day northern Tunisia, northeastern Algeria, and western Libya).[1] The region was one of the earliest to come under Roman control following the destruction of Carthage in 146 BC. However, little of the surviving Roman material in North Africa dates from before the late first century AD, and Rome's influence on the local culture was not dominant until the second century AD, when the area became a primary source of grain for Italy and a source of wealth for aristocratic Romans. Reflecting this increasing affluence, a growing number of public buildings and private residences were adorned with ornate mosaic floors, and from the end of the second century AD onward, mosaics were a standard form of decoration in wealthy Roman villas.

The earliest Roman mosaics in North Africa made use of elaborate black-and-white patterns and geometric designs prevalent in Italy during the first and second centuries AD. Mosaics of this type were probably introduced to North Africa by itinerant craftsmen who had established local workshops along the east coast of Africa Proconsularis by the early part of the second century AD. The impact of the Italian tradition is evident in the earliest mosaics at the prominent centers of Hadrumetum (present-day Sousse) and Thysdrus (present-day El Djem), which display black-and-white geometric designs unparalleled in previous mosaic production in North Africa.[2] The local mosaic styles in North Africa, however, soon diverged from contemporary Italian trends by adding stylized vegetal elements and using vibrant color for both figural and decorative compositions. Elaborate, polychrome mosaics similar in composition to that of the black-and-white Getty Medusa mosaic (cat. 1), for example, survive from a villa at Thysdrus and from a bath complex at nearby Dar Zmela.[3] This rapid transformation was likely influenced by the abundant supply of colored limestone and marble that was available locally. It is also clear from the decoration of buildings such as the Great Baths at Thysdrus, however, that mosaic floors in the black-and-white style continued to be produced alongside the more elaborate polychrome pavements well into the second century AD.[4]

Mosaics of a very different tradition also existed in North Africa during the same period, although examples of these are fewer. Detailed polychrome *emblemata* depicting a wide variety of figural subjects appear as individual scenes within the designs of larger floors. These picture panels were often set in a decorative geometric field or distributed around a primary scene that was associated with mythological figures, combats in the amphitheater, or the hunt. The Getty's Mosaic of a Lion Attacking an Onager (cat. 5), which depicts a single scene surrounded by a *guilloche* border, belongs to this tradition. Its original context may have been comparable to that of a pair of *emblemata* from the House of the Dionysiac Procession at Thysdrus, which were set at either end of a frieze at the entrance to the *triclinium*; the room itself was decorated with an elaborate mosaic floor composed of intricate vegetal designs.[5] *Emblemata* were also displayed together in groups, as in the Calendar Mosaic with the Seasons and the Months in the House of the Months at Thysdrus, which was composed of sixteen individual panels set into a floral-style frame.[6] The Getty mosaic may have been arranged similarly as one of a series of related panels, perhaps also illustrating scenes of hunting or the arena.

Over the course of the third and fourth centuries AD, pictorial mosaics in North Africa increasingly displayed a preference for large-scale figural compositions.[7] Themes of the amphitheater and the hunt were especially widespread. In no other regions of the Roman Empire are they found in such variety and abundance. Some of the earliest examples include early third-century AD hunting mosaics from Carthage and from Oudna, as well as a mosaic of similar date depicting amphitheater combats from Thysdrus and a hunting pavement from El Kef in western Tunisia that may date to the late second century AD.[8] Both the composition and the imagery of North African mosaics during this period influenced those of other regions of the Roman Empire, particularly Sicily and Italy. One such example is the Getty's fourth-century AD Mosaic Floor with Bear Hunt (cat. 2) from a Roman villa near Baiae.[9] The influence of North Africa's large-scale narrative mosaics on the eastern part of the empire is also evident. It is clearly demonstrated by the mid-fifth-century AD Mosaic of Megalopsychia from Antioch, in which a personification of Megalopsychia (Generosity) is surrounded by a series of *venationes* and animal combats, and by the early sixth-century AD Mosaic of the Worcester Hunt, also from Antioch, which depicts various hunting scenes surrounding a central figure.[10]

NOTES

1. For a comprehensive examination of Roman mosaics in North Africa, see Dunbabin 1978. Very different geographic and historical conditions affected the development of mosaic production in the mountainous provinces of Numidia and Mauretania, which retained separate regional and local traditions.
2. Examples include House B of the Terrain Jilani Guirat (Foucher 1960a, 44–46, plate 8) and the House of the Peacock (Maison du Paon) (Foucher 1961, 3–14, plates 1–5) at Thysdrus, both dating to the early second century AD.
3. Thysdrus (a villa): Foucher 1963, 97, fig. 13b. Dar Zmela (a bath complex): Foucher 1960b, 121–22, no. 57.247; and Foucher 1963, 97, fig. 13a.
4. For the baths at Thysdrus, see Gauckler 1910, 61; and Foucher 1960a, 103.
5. Foucher 1963, 90–96, figs. 11d–e.
6. Foucher 1961, 30–50, plates 32–34; and Parrish 1984, 156–60, no. 29, plates 42–44.
7. Lavin 1963, 229–42.
8. Oudna: Gauckler 1910, 122–23, no. 362; and Lavin 1963, 230–31, fig. 75. Carthage: Poinssot and Lantier 1923, 154–58; and Lavin 1963, 233, fig. 79. Thysdrus: Merlin 1915, no. 71f, 4; and Lavin 1963, 231–32, fig. 77. El Kef (or Le Kef): Lavin 1963, 231–32, fig. 76. For another hunting mosaic from Thysdrus: Gauckler 1910, 26, no. 64; and Lavin 1963, fig. 80.
9. For examples with references, see the introduction to Italy in the present catalogue. For additional connections reflected in the mosaics of Sicily and North Africa, see Wilson 1982, 413–28.
10. Mosaic of Megalopsychia (Hatay Archaeology Museum, 1016): Lassus 1934, 114–56, figs. 1–27; Levi 1947, 326–45, fig. 136, and plates 76b–80; and Lavin 1963, 189–90, figs. 6–7. Mosaic of the Worcester Hunt (Worcester Art Museum, 1936.30–31): Levi 1947, 363–65, plate 90a; Lavin 1963, 187–90, fig. 2; and Becker and Kondoleon 2005, 228–37, no. 8, dated to AD 480–520.

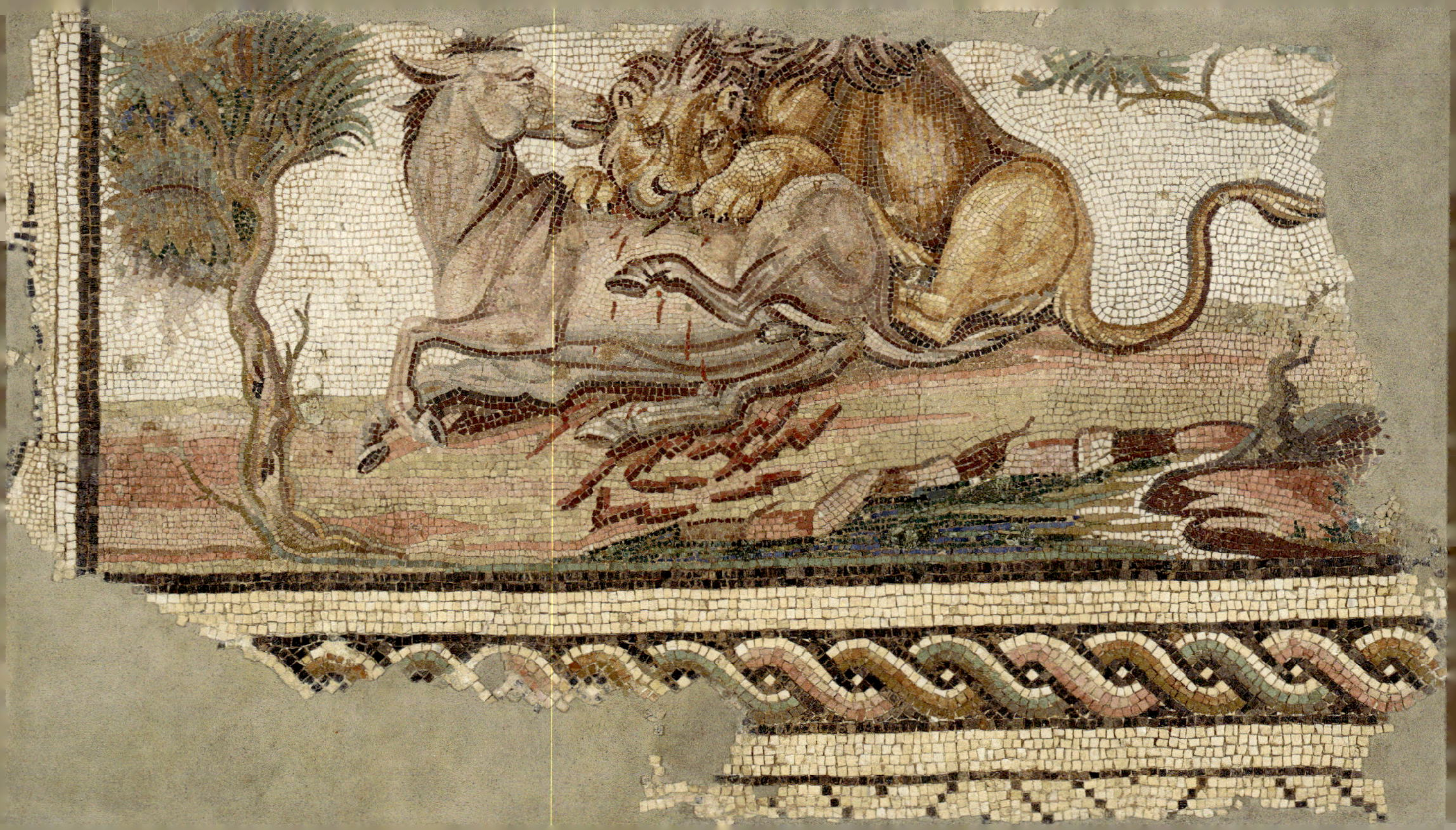

5. Mosaic of a Lion Attacking an Onager

Roman, from Hadrumetum (present-day Sousse), Tunisia, AD 150–200
Stone and glass tesserae, 98.4 cm × 160 cm
73.AH.75

Provenance

This mosaic panel was discovered near Hadrumetum (present-day Sousse), on the east coast of Tunisia, but the exact findspot is uncertain. It was excavated sometime before 1914.[1] The J. Paul Getty Museum purchased the mosaic in 1973.[2]

Commentary

A ferocious lion attacks an onager, or wild ass, sinking its teeth and claws into the victim's back and forcing the animal to the ground. The lion faces outward, directly engaging the viewer with its gaze, while the terrified onager struggles, twisting its head backward. This violent event is set in a landscape framed by trees. Blood from the wounded animal flows on the ground toward the bank of a spring or a stream. A *guilloche* pattern decorates the remains of the mosaic's lower border.

The image of a lion tearing into the back of its captured prey was pervasive in mosaics of the Roman Empire. The motif had a long history in Near Eastern and Greek art, dating back to the Bronze Age, but immediate antecedents can be found in the fourth and third centuries BC in several Greek pebble mosaics, in which similar compositions appear as part of a decorative border. A mosaic floor of the early fourth century BC from the House of the Mosaics at Eretria depicts a clash between a lion and a horse as part of a larger frieze including griffins and mythical human figures.[3] Similarly, the decorative frieze of a mosaic from the Pompeion in Athens shows lions attacking animals in the corners and heraldic griffins reclining along the sides.[4] In other examples, individual scenes of animals in combat were enlarged and placed in separate panels of the mosaic, giving them greater prominence. Pavements from a villa in Sparta and from the peristyle of the House of the Mosaics at Motya (present-day Mozia) in Sicily display framed scenes of lions attacking bulls.[5] The example from Motya is placed next to an identical composition of a griffin with a horse.[6] However, in contrast with the Roman version of the scene represented in the Getty mosaic, there is no indication of a landscape setting in these earlier works.

In Roman mosaics, the motif of animal combats became closely associated with themes of the amphitheater and the hunt, which were immensely popular forms of public entertainment involving wild animals and humans.[7] At the same time, images of lions or other wild cats overpowering their fallen prey from behind, often depicted in a landscape setting, frequently appear as individual *emblemata*. The representations of these ferocious animals on their own, facing outward to confront the viewer, may have served an apotropaic function—warding off evil—rather than a narrative one. The original context of the Getty mosaic is unknown, but it was likely an *emblema* in a much larger composition that included additional scenes decorating the floor of a wealthy Roman villa.

Comparanda

Elements of the Getty mosaic resemble two relatively small mosaic panels that formed part of the floor decoration of the Imperial Palace of Hadrian's Villa at Tivoli, near Rome, dated about AD 120–130. In one example, from the Basilica of the Palace, a lion attacks a bull in a rugged landscape, while another bull observes in the background.[8] The second mosaic, which originally adorned the main *triclinium*, depicts a battle between wild cats and centaurs, one of whom has been taken down by a tiger.[9] The exceptionally detailed composition, naturalistic rendering of the scenery, and painterly style of these mosaics follow earlier Hellenistic Greek models, possibly copying a painting or *emblema* from that period.[10] In comparison, the Getty mosaic, although similar to the pavements from Hadrian's Villa in theme and composition, exhibits shallow perspective and a limited landscape that is more characteristic of Roman mosaics at the end of the second century AD. The Getty panel may have belonged to a series of individual mosaics depicting pairs of fighting animals, as seen in the decoration of the atrium of the third-century AD Villa of the Laberii at Oudna in Tunisia. In this instance, a fragmentary mosaic panel showing a female lion overpowering an onager was one in a series placed between the columns of the courtyard.[11]

Other mosaics with *emblemata* comparable to the Getty mosaic feature central mythological scenes of deities, most notably Dionysos, tamer of wild beasts and patron god of the Telegenii, an association in North Africa that participated in the organization of animal spectacles for the arena. A mosaic floor of the mid-second century AD that decorated the *triclinium* of the House of the Dionysiac Procession at Thysdrus (present-day El Djem) includes a pair of similar panels depicting animal attacks, framed by a *guilloche* border.[12] These were placed at either end of a processional scene of the child Dionysos riding a submissive lion. One mosaic panel depicts a tiger and two onagers in a rocky landscape with trees and is similar in composition to the mosaic with two bulls from Hadrian's Villa but closer in style to the Getty panel. The tiger attacks one onager while the second ass flees. The companion piece shows two lions on either side of a boar, also in a landscape setting. Like the lion in the Getty mosaic, the tiger and one of the lions stare straight out at the viewer. The expressive features of the lion in the Getty mosaic may place it closer in date to a similar representation of the scene on a fragmentary panel from Carthage dating to the mid-third century AD, now in the Smithsonian Institution National Museum of Natural History in Washington, DC.[13] This image, which also depicts a lion and an onager, is one of several animal scenes that once surrounded a central female figure riding in a chariot drawn by stags.[14] The figure is probably Artemis the huntress, whose associated imagery, like that of Dionysos, includes wild animals.

Condition

Only part of the left and bottom edges of the border survive. Some sections of the mosaic may have been restored in ancient times.[15]

Bibliography

Foucher 1960b, 17, no. 57.040, plate 7a; *Le bien public*, Dijon, April 19, 1961; Foucher 1963, 90, fig. 11c; Balil 1964, 7–8, no. 57.040, fig. 3; Sotheby Parke Bernet, New York, May 4, 1973, lot 199; Neuerburg 1975, 51; Parrish 1987, 113–34.

A.B.

NOTES

1. According to Foucher, the mosaic was found prior to 1914 near the avenue Tahar-Sfar (formerly avenue Marechal-Fochs) and was then in the possession of the family of the local fire chief; see Foucher 1960b, 17, no. 57.040, plate 7a. Neuerburg, however, writes that it was found at Oued Kharroub, near Sousse, citing *Le bien public*, Dijon, April 19, 1961; see Neuerburg 1975, 51.
2. Sotheby Parke Bernet, New York, May 4, 1973, lot 199.
3. Ducrey, Metzger, and Reber 1993, 85–96; and Salzmann 1982, 90–91, no. 37, pl. 26.
4. Salzmann 1982, 86, no. 19, plate 24. A similar example from Corinth can be found in Salzmann 1982, 95, no. 64, plate 23.
5. The mosaic from Sparta, now in the Archaeological Museum of Sparta, is slated to be published by the excavator, E. Kourinou; see Panayotopoulou 1998, 112. For the mosaic from Motya, see Tusa 1997; and Famà 1997.
6. In Greek mosaics, the lion motif is commonly accompanied by or interchanged with griffins. A mosaic from an early fourth-century BC house at Olynthus includes a scene with two griffins, in place of lions, leaping upon a stag from either side and sinking their beaks and claws into its back; see Salzmann 1982, 99, no. 78, plate 13.
7. Parrish 1987, 128–34, provides examples with bibliography for the mosaics in the amphitheater at El Djem, Nennig (in present-day Germany), and at Antioch, and for hunting scenes at Ain Tounga in Africa Proconsularis, Piazza Armerina in Sicily, and the Great Palace at Constantinople.
8. Aurigemma 1961, 169–70, plate 16.
9. Schwarzmaier, Scholl, and Maischberger 2012, 232–33, no. 131, Staatliche Museen zu Berlin, Die Antikensammlung.
10. Although it is generally thought that these mosaics represent the highest quality in Roman mosaic work, it has also been argued that they were Hellenistic originals. See the comments by Christine Kondoleon on the mosaics as Hellenistic *emblemata* produced in the first century BC and preserved as heirlooms in the introduction to the present volume.
11. Gauckler 1897, 185, fig. 1 (*Atrium* 30, 2), 195, fig. 4, and plate 20; Gauckler 1910, no. 370; and Parrish 1987, 119, fig. 9.
12. Foucher 1963, 90–96, figs. 11d–e.
13. Smithsonian Institution National Museum of Natural History, acc. no. 012416. Parrish 1987, 117, fig. 7; and Heap 1883, 415–16.
14. For a description of the complete mosaic, now lost, see Heap 1883, 416.
15. Neuerburg 1975, 51.

SYRIA

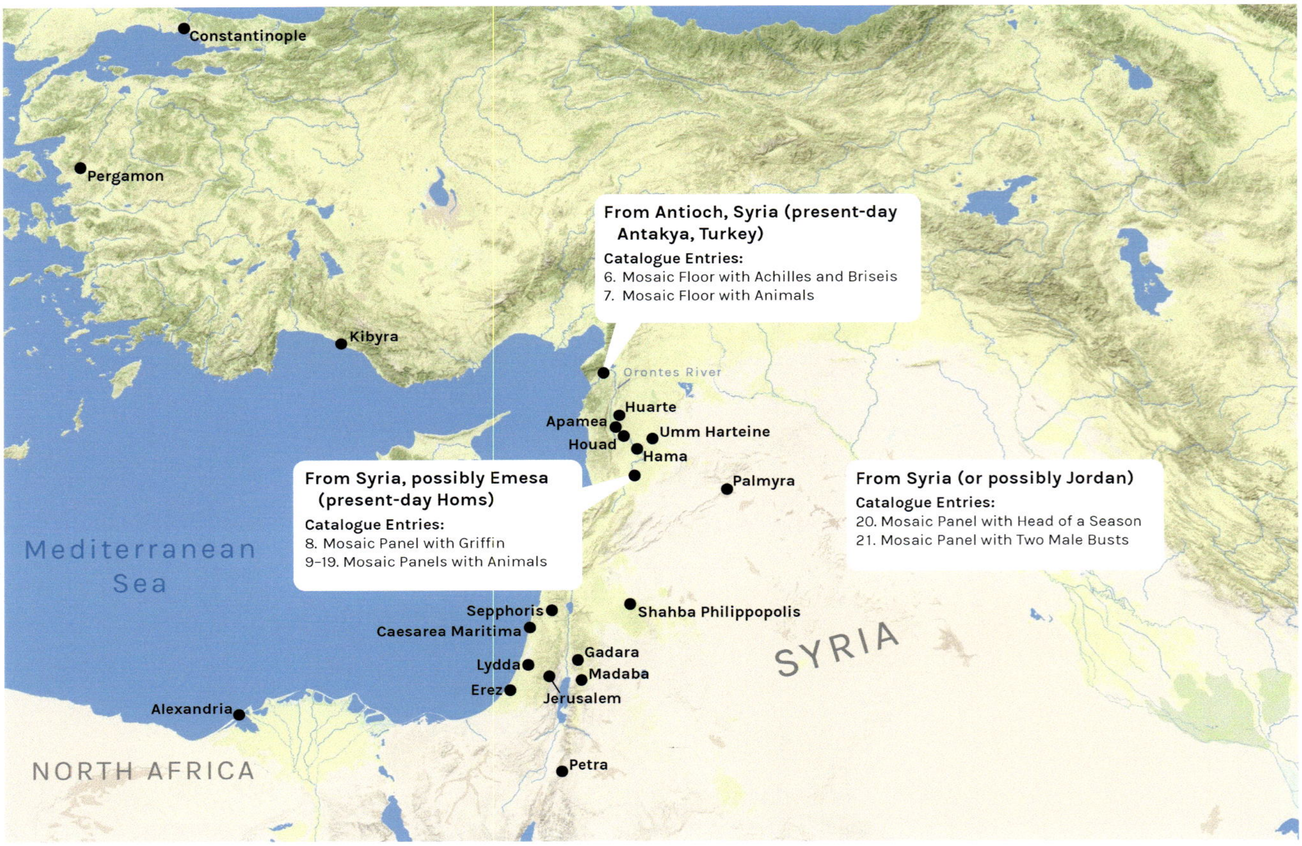

Syria

Although few mosaic pavements in Syria dating before the first century AD survive, there can be little doubt that mosaic production in the region, one of the wealthiest provinces of the Eastern Roman Empire and formerly the center of the Greek Seleucid kingdom, evolved directly from Hellenistic Greek traditions. Unlike Gaul and North Africa in the first and second centuries AD, where the style of mosaics was strongly influenced by the presence of itinerant craftsmen who introduced the black-and-white style popular in Italy at that time, Syria preferred the Hellenistic style of detailed narrative mosaics that imitated paintings. The abundance of Roman mosaics preserved at various sites, notably at the great metropolis of Antioch (present-day Antakya, Turkey), demonstrates that the pictorial style remained dominant in Syria until at least the end of the fourth century AD.[1]

The mosaics excavated at Antioch and the surrounding area date from about the beginning of the second century AD until soon after the destruction of the city by earthquakes in AD 526 and 528.[2] Figural scenes from classical mythology, often inspired by literary sources, were especially prevalent. The Atrium House at Antioch illustrates the characteristic style of mosaics from the region in the early second century AD. Three separate mythological scenes decorate the pavement of the *triclinium*: the love story of Aphrodite and Adonis; the Judgment of Paris; and a drinking contest between Dionysos and Herakles.[3] The naturalistic treatment of the figures, the rendering of three-dimensional space, and the variety of colors reflect the influence of Hellenistic painting. Although the origin of the Getty's Mosaic Floor with Achilles and Briseis (cat. 6) is unknown, the style of the mosaic, as well as its conception as a framed picture panel, is clearly part of this tradition. The use of geometric patterns and vegetal motifs as borders surrounding figural panels is reminiscent of the framing of Hellenistic *emblemata*. Additionally, while few mosaics are preserved in other parts of Syria before the third century AD, the cities that later became prominent production centers, such as Apamea, Emesa (present-day Homs), and Shahba Philippopolis, continued to draw on a similar style of pictorial composition.

The influence of Hellenistic Greek art and a preference for subjects from Greek mythology characterize Syrian mosaics well into the Christian period, long after different styles and themes had become more popular in other regions of the Roman Empire. At the end of the fourth and beginning of the fifth centuries AD, however, the use of complex narrative themes diminished in favor of simpler figural compositions framed by elaborate decorative patterns.[4] A distinctive ornamental design that had been introduced in mosaics of the Roman East during the second century AD—the so-called rainbow style, in which colored tesserae are arranged in a diagonal sequence to produce a kaleidoscopic effect—became increasingly widespread in the mid-fourth century AD. The spiral pattern of alternating colors surrounding the central figural panel (with a hare eating grapes) of the Getty's Mosaic Floor with Animals (cat. 7), from the Bath of Apolausis at Antioch, exemplifies this style. Such designs were not confined to secular settings; mosaic pavements with various versions of the rainbow style cover entire floors of a church in Kaoussie, a suburb of Antioch, which is dated by an inscription to AD 387.[5]

By the fifth century AD, a new repertoire of images developed to fit the needs of the growing number of churches established throughout the region. Animals, both real and fantastic, dominate the imagery of fifth- and sixth-century Syrian church mosaics. Arranged freely across the floor, they are typically shown walking or standing among small landscape elements. They are occasionally paired in lively chase scenes, such as the Getty's mosaic panel depicting a lion pursuing a bull (cat. 19). Vegetal components, such as scrolling vines, were placed around an assortment of animals—lions, bulls, and peacocks and other birds—alluding to the Christian vision of Paradise. The majority of the beasts represented on the Getty's mosaic panels with animals (see cats. 9–19), which are thought to be from an early Christian church in Homs, can be read simply as representing the great variety of creation.

NOTES

1. A pebble mosaic from Tarsus in Cilicia (in present-day Turkey) dates to the late third or early second century BC (Salzmann 1982, 37, 113, no. 125); and mosaics in Palestine have been dated to the late second or first century BC (Dunbabin 1999, 187–88nn1–3). On Hellenistic mosaics in general, see Westgate 2002.
2. On the mosaics of Antioch and its vicinity, in general, see Levi 1947; Campbell 1988; Cimok 2000; Kondoleon 2000; and Becker and Kondoleon 2005.
3. Levi 1947, 15–25, plates 1–2a, 142a, 145–48. The mosaic of Aphrodite and Adonis is now in the Princeton University Art Museum (y1940-156); the Judgment of Paris is in the Louvre (Ma 3443); and the drinking contest between Dionysos and Herakles is in the Worcester Art Museum (1933.36).
4. Kitzinger 1965; Balty 1984; and Donceel-Voûte 1994.
5. Levi 1947, 283–85, 423–25 (on the decorative style of the mosaics), plates 113–15, 139; and Donceel-Voûte 1988, 21–31, with additional bibliography.

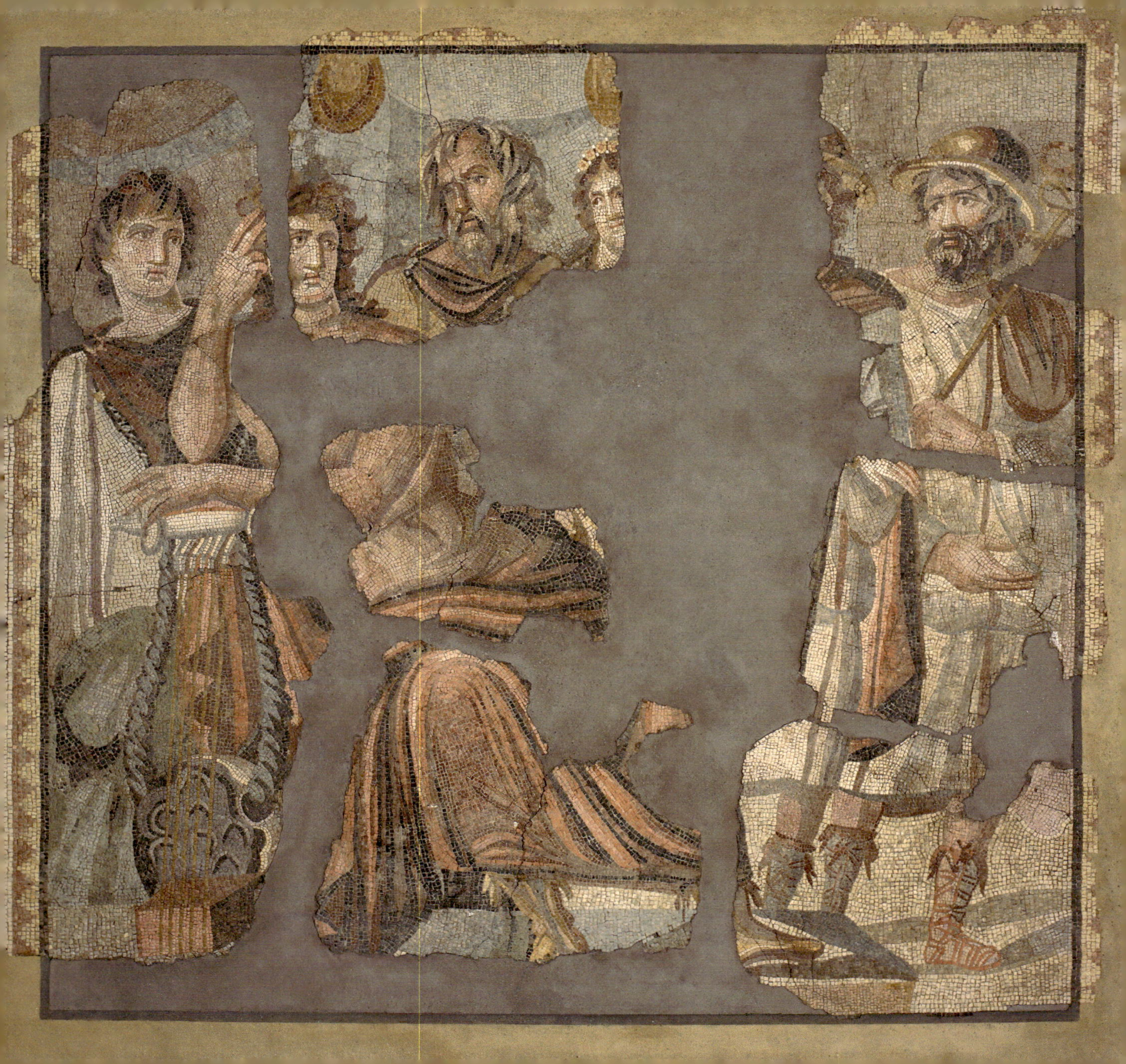

6. Mosaic Floor with Achilles and Briseis

Roman, from Syria, possibly Antioch (present-day Antakya, Turkey), AD 100–300
Stone and glass tesserae, 231.1 cm × 240 cm
68.AH.12

Provenance

Although the exact origin of this mosaic is unknown, it was most likely discovered somewhere in the Roman province of Syria, probably in the vicinity of Antioch (present-day Antakya, Turkey). It was purchased by the J. Paul Getty Museum in 1968.[1]

Commentary

A dramatic passage from Homer's epic poem the *Iliad* (1.327–48) inspired the scene depicted on this mosaic: the dispute between the Greek warrior Achilles and King Agamemnon over the concubine Briseis, whom Achilles had taken prisoner during the Trojan War. Persuaded by Athena, Achilles reluctantly gives up Briseis but then angrily refuses to carry on fighting the war. He returns to battle only later, after the death of his companion Patroclus. Six figures are portrayed in the version of the episode represented on this mosaic. A youthful male wearing a *chlamys* (cloak), probably Patroclus, stands on the far left. Achilles sits beside him holding a lyre and leaning his head on his right hand in a gesture of grief and resignation. A large portion of his upper body is missing, as well as part of his boots. An elderly bearded man standing behind Achilles's left shoulder may be Phoenix, who accompanied Achilles on the expedition against Troy. The fragmentary female figure next to him is Briseis. Only a portion of her head is preserved. On her left, Eurybates and Talthybios, the two heralds who will take her to Agamemnon, observe the scene. Large sections of the figure closest to Briseis are missing. The final figure, essentially complete, is bearded, holds a *kerykeion* (a herald's staff with intertwining snakes), or caduceus, and wears the characteristic broad hat of a herald. Two shields in the background hold up a curtain, representing the tent in which the exchange took place.

The life of Achilles was a popular theme in Roman art, which most often celebrated episodes from his childhood or the occasion when Odysseus discovers Achilles on the island of Skyros and persuades him to join the Trojan War. The story of Achilles and Briseis was not frequently depicted; the few surviving representations appear in a variety of different media and time periods. Although versions of the scene appear in Greek vase painting by the early fifth century BC, the earliest parallel to the Getty mosaic in style and composition is a first-century AD wall painting from the atrium of the House of the Tragic Poet in Pompeii.[2] Like the Getty mosaic, this fresco depicts a series of figures on seemingly different planes to create the impression of depth. In the foreground of the Pompeian fresco, Achilles sits next to Patroclus, who turns his back as he seizes Briseis's wrist. The two heralds along the left side of the scene appear to be slightly behind the central group; Phoenix stands behind Achilles's chair, his hand resting upon it. In the far background, beyond a row of soldiers, an opening in the tent affords a view of the sea. The fresco is thought to be a copy of a Hellenistic painting in view of its detailed composition, the careful rendering of space, and the expressive features of the characters.

Some of the most complete representations of the episode of Achilles and Briseis survive in late Roman metalwork, including a bronze sheath and a silver *missorium* of the fourth century AD, as well as an engraved bronze *situla* dating to the fifth century AD.[3] The scene also appears in other examples from late antiquity, such as the relief decoration of a stone architrave dating from the fourth or fifth century AD, a fourth-century AD papyrus, and the painted miniatures of the *Ilias Ambrosiana*, a late fifth- or early sixth-century AD illuminated manuscript of the *Iliad* on vellum.[4] Although these works are based on the same literary account, the choice and arrangement of the figures vary.

Comparanda

The scene of Achilles and Briseis as it appears on the Getty mosaic is found on only three other mosaics, all dating to the second and third centuries AD. Two of these were discovered at Antioch. The mosaic in the House of Briseis's Farewell, while very fragmentary, most closely resembles the Getty mosaic in its overall composition, with two heralds standing on the right—one bearded, wearing the herald's hat and also holding a *kerykeion*—and Patroclus on the opposite side, holding Briseis's hand. Achilles appeared in the part of the scene that is now lost.[5] The other mosaic discovered at Antioch, from the House of Aion, depicts only three figures, each identified by an inscription: *Briseis*, *Achilles*, and *Talthybios*.[6] The final mosaic, the latest of the three, dating to the late third century AD, is from a fragmentary floor of a house in Sparta; it is similarly inscribed with the names *Briseis*, *Achilles*, and *Talthybios*.[7]

The Getty Museum's mosaic and the mosaic from the House of Briseis's Farewell at Antioch demonstrate a remarkable continuity with Hellenistic artistic tradition, particularly in the naturalistic rendering of the figures and the attention to detail.[8] The similarities in style and composition of these two mosaics support a contemporary date, most likely in either the second or the third century AD, and suggest that both are the product of a workshop in Antioch or its immediate vicinity.[9] In addition, the theme, an uncommon choice for mosaics outside the Roman East, may also reflect a regional preference.

Condition

This mosaic underwent extensive restoration prior to its arrival at the J. Paul Getty Museum. Three fragmentary panels of the mosaic had been attached to each other to form one piece, and large sections had been filled in (using AJK dough, an alternative to plaster) to complete the square form of the mosaic. In 1998, Getty conservators disassembled and cleaned the mosaic before mounting it on an aluminum panel.

Bibliography

Spink & Son, Ltd. 1966, 7; *London News* 1966, 47; Vermeule and Neuerburg 1973, 51, no. 110; von Gonzenbach 1975, 401–8, plate 194; Fredericksen, Frel, and Wilson 1980, 41; Balty 1981, 365; *LIMC* s.v. "Briseis," 158–59, no. 6.

A.B.

NOTES

1. Purchased from Spink & Son, Ltd., in London, who had acquired the mosaic by 1966 (Spink & Son, Ltd. 1966, 7).
2. The scene appears on a red-figure vase from Vulci in the British Museum (E 76) (*LIMC* s.v. "Briseis," 158, no. 1; *ARV*² 406, 1) and an Athenian red-figure *skyphos*, painted by Makron, in the Louvre (G 146) (*LIMC* s.v. "Briseis," 158, no. 2; *ARV*² 458, 2). The wall painting from Pompeii is now in the Museo Archeologico Nazionale in Naples (9105) (*LIMC* s.v. "Briseis," 158, no. 3; and von Gonzenbach 1975, plate 193).
3. The bronze sheath is in the British Museum (1772,0303.12) (*LIMC* s.v. "Briseis," 159, no. 9). The *missorium* is in the Cabinet des Médailles, at the Bibliothèque Nationale de France, in Paris (BnF 56.344) (*LIMC* s.v. "Briseis," 159, no. 8; and Leader-Newby 2014, 91–95). This *missorium*, traditionally known as the "Shield of Scipio," depicts a grouping of figures similar to that of the Getty mosaic, although the subject is generally interpreted as conflating the episode of Briseis with that of the embassy to Achilles seeking his return to battle. The *situla*, formerly in the Doria collection, Rome, is now missing (*LIMC* s.v. "Briseis," 159, no. 10; and Bianchi Bandinelli 1955, fig. 26).
4. The architrave is now in the Coptic Museum in Cairo (*LIMC* s.v. "Briseis," 159–60, no. 11). For the papyrus, now in the Bayerische Staatsbibliothek in Munich (Papyr. gr. 128), see Hartmann 1930, plate 17; and *LIMC* s.v. "Briseis," 160, no. 13. For the *Ilias Ambrosiana*, in Milan, Biblioteca Ambrosiana Cod. 1019 (formerly F 205 Inf.), see Bianchi Bandinelli 1955, Min. VI, 55, figs. 42 and 105; and *LIMC* s.v. "Briseis," 160, no. 12.
5. Levi 1947, 46–49, plate 8a; and *LIMC* s.v. "Briseis," 158, no. 4.
6. Levi 1947, 196–97, plate 43c; and *LIMC* s.v. "Briseis," 158, no. 5.
7. Christou 1964, 137–38 (Room 2); and *LIMC* s.v. "Briseis," 159, no. 7.
8. For a discussion of the general characteristics of mosaics from Antioch, see Kondoleon 2000.
9. The general stylistic characteristics seem to place the mosaic in the second century AD; see Balty 1981, 365; von Gonzenbach 1975, 401–8; and Levi 1947, 48–49. Campbell compares the modeling of the faces and the weight of the drapery with mosaics in the House of the Drinking Contest at Seleucia, also dated to this period; see Campbell 1988, IV A 21; and Levi 1947, 156, esp. plates 30b and 32d. However, a later date, in the third or even the early fourth century AD, has been proposed by Katherine Dunbabin; see the 1984 consultant report in the files of the antiquities department of the J. Paul Getty Museum. Regarding the origin of the Getty mosaic, Balty suggests a workshop at Antioch, observing similarities between the heads of Achilles and Patroclus in the Getty mosaic and the bust of Spring in the Calendar Mosaic; see Balty 1981, 365; and Levi 1947, 36–38, plate 5b, and between the head of Phoenix and that of Oceanus in a second mosaic from the same house; see Levi 1947, 38–39, plates 6, 149a, both dating to the second century AD.

Antioch and the Bath of Apolausis
History of the Excavations

Founded in 300 BC, Antioch (present-day Antakya, Turkey) was one of the most important political and cultural centers of the Greek East, and it became one of the great metropolises of the Roman Empire. Extensively excavated, the ancient site is recognized in particular for its widespread use of elaborate mosaic decoration, which adorned private houses, public buildings, and churches. During excavations in the 1930s at Antioch, its wealthy suburb Daphne, and the port city of Seleucia Pieria, archaeologists unearthed more than three hundred mosaic pavements.[1] While the majority of the mosaics decorated private villas, mosaics also embellished public buildings such as bath complexes, which Antioch was renowned for in antiquity. Libanius and other ancient authors celebrated the clear and abundant waters of the city, as well as the great number and splendor of public and private baths, fountains, and *nymphaea*. The excavations at Antioch brought to light the long tradition of decorating such structures with impressive paintings and mosaics.

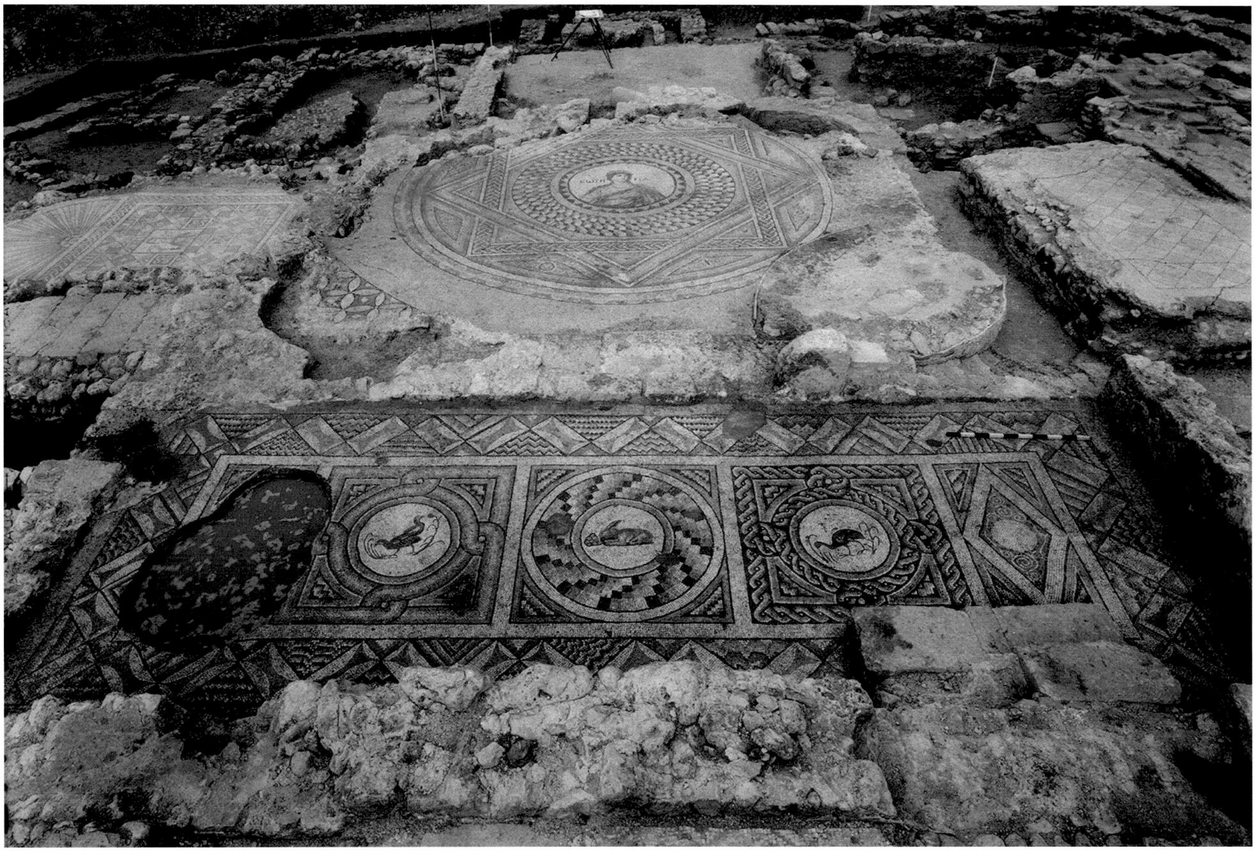

Figure 17. Excavation photo showing Mosaic Floor with Animals from the Bath of Apolausis, Antioch, Syria, 1938. Antioch Expedition Archives, Department of Art and Archaeology, Princeton University, no. 4092

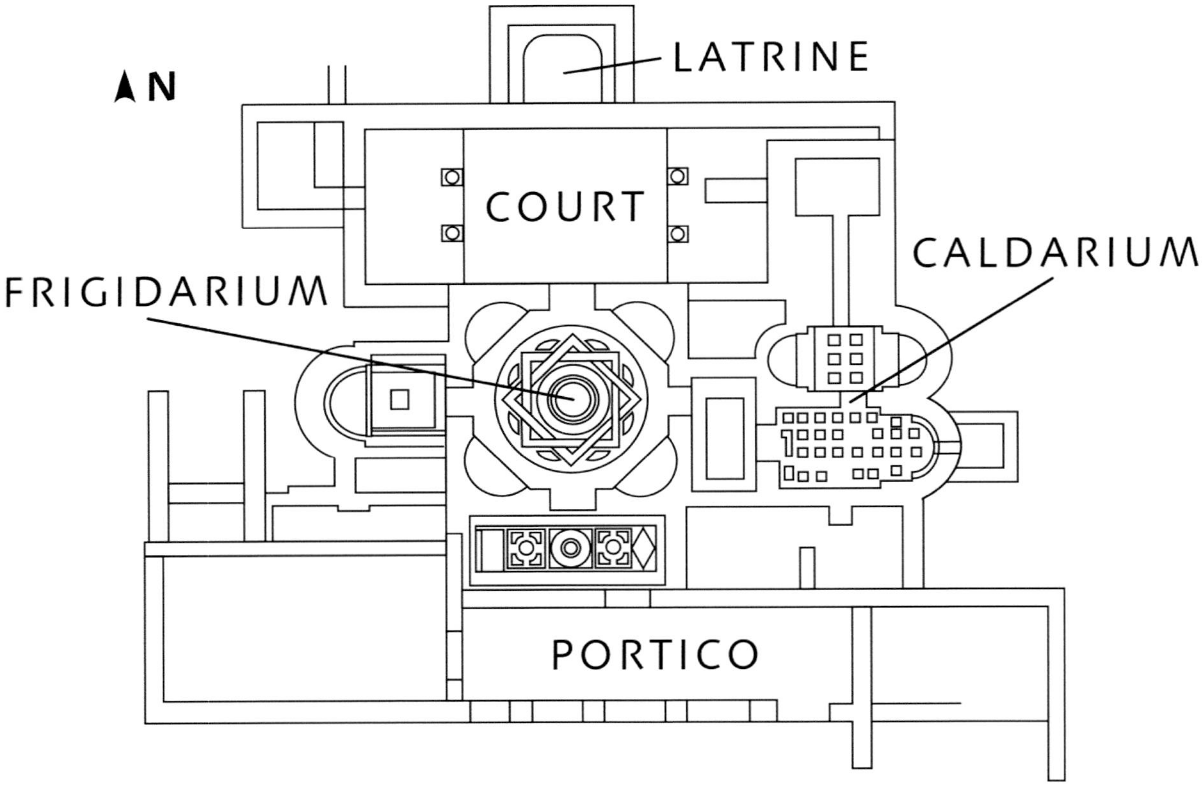

Figure 18. Plan of the Bath of Apolausis, Antioch, Syria, based on an original excavation drawing (see Stillwell 1941, plan 5)

The Getty Museum's Mosaic Floor with Animals (cat. 7) was uncovered during the excavation of the Bath of Apolausis, a small Roman bath discovered on the eastern side of the plain of Antioch, at the foot of Mount Silpios.[2] The bath, a public building that originally served an agricultural complex or group of country villas, was richly decorated with floor mosaics and wall frescoes. Mosaics were used primarily in the decoration of the main rooms of the bath. The Getty's mosaic paved a vestibule connecting the entrance to the bath—a portico along the south side of the building—with the largest room—a central octagonal space with large niches that functioned as both the *frigidarium* and the main social hall (figs. 17, 18). Geometric patterns decorated the floor of the portico while a mosaic depicting a female bust of Soteria, the Greek personification of Salvation, covered the floor of the *frigidarium* (fig. 19).[3] A second figured mosaic—a personification of Apolausis (Enjoyment), after which the bath was named—decorated the bottom of a large pool with an apsidal end accessed through a doorway on the west side of the octagonal hall (fig. 20).[4] At the time they were excavated, the elaborate floor mosaics decorating the bath were relatively well preserved, but the only significant remains of fresco decoration were found in the vestibule with the Getty mosaic. On the walls of this room, traces of a painted revetment imitated marble slabs, which were arranged in alternating contrasting colors: white veined with blue, black, shades of red, and yellow.[5]

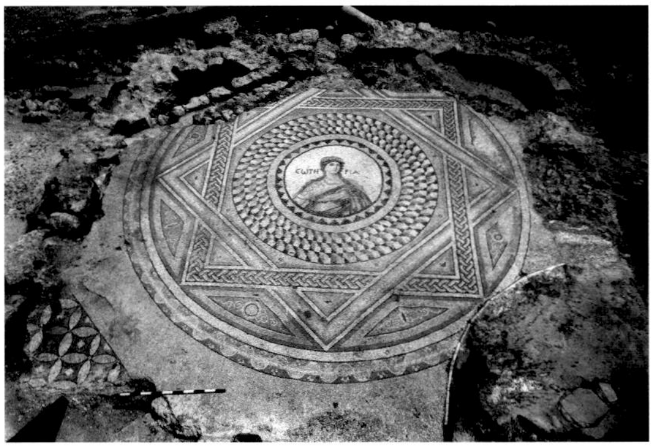

Figure 19. Excavation photo showing the mosaic of Soteria from the Bath of Apolausis, Antioch, Syria, 1938. Antioch Expedition Archives, Department of Art and Archaeology, Princeton University, no. 3955

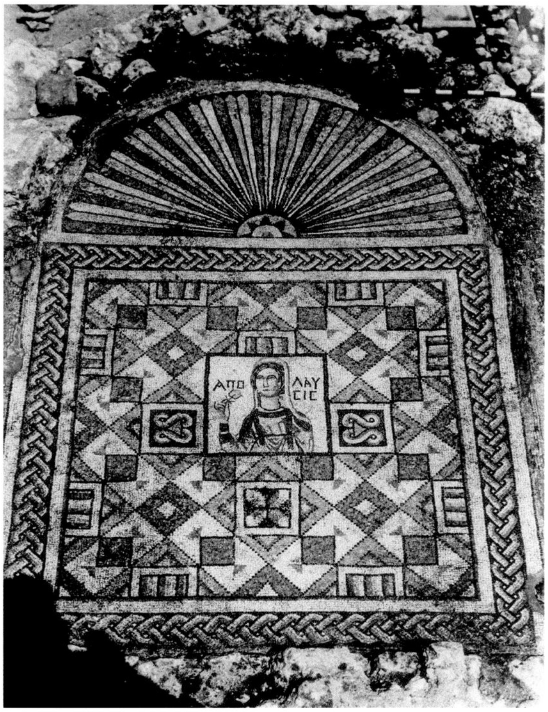

Figure 20. Excavation photo showing the mosaic of Apolausis from the Bath of Apolausis, Antioch, Syria, 1938. Antioch Expedition Archives, Department of Art and Archaeology, Princeton University, no. 4082

To the east of the central hall, an opening led to the antechamber of a *caldarium*. This antechamber, paved with a mosaic of geometric patterns and scrolling vines, had an entrance at either end of the room. The *caldarium* consisted of two apsidal rooms that were heated by a hypocaust system, in which a raised floor supported by *pilae* (pillars) created channels underneath the building to distribute warmth from the furnace. The *pilae* of the hypocaust in the Bath of Apolausis stood on a tiled floor, and a square chamber at the end of a passageway to the north housed the furnace. The northern section of the bath, consisting of a large courtyard flanked by two-columned porticoes and a latrine, was not decorated with mosaics.

The bath itself occupied the northern end of a substantial architectural complex where fewer remains have been preserved. Buildings to the east and the south were partially covered by a modern farm, which made extensive use of architectural fragments from the ancient structures. No additional mosaics were discovered. Though relatively large, these buildings were unremarkable in comparison to the high-quality construction and the rich decoration of the bath. While some of the ancient remains reused in the modern walls have been dated to as early as the second or third century AD, the excavations of the bath and the surrounding buildings suggest that the main period of use was in the fourth and fifth centuries. Although there is evidence for construction in the mid-fifth century, the parts of the bath dating to this period are thought to be restorations or additions to an original structure built in the late fourth century AD.[6] Based on stylistic comparisons with other mosaic pavements from Antioch and the surrounding region, it has been determined that the Getty mosaic should also date to the very end of the fourth century.[7]

A.B.

NOTES

1. Eight campaigns were undertaken from 1932 to 1939. For a catalogue of the mosaics, see Stillwell 1941, 171–219; and Levi 1947.
2. The site was located near present-day Toprak-en-Narlidja. On the excavations of the Bath of Apolausis, see Stillwell 1941, 19–23; and Levi 1971, 304–6.
3. The mosaic of Soteria is now in the Hatay Archaeological Museum in Antakya, Turkey (inv. no. 977, excavation no. 5287-M82A).
4. The Apolausis mosaic is at Dumbarton Oaks Research Library and Collection (BZ.138.72).
5. The surviving frescoes were extremely fragmentary. Stillwell notes that the French excavator, Jean Lassus, described traces of bright red paint on the exterior of the building; see Stillwell 1941, 22.
6. See Levi 1947, 304, 626.
7. Balty supports Levi's date on this basis, citing in particular the Church of Khirbet Muqa in Apamea (AD 394/395); see Balty 1995, 92, and especially 89–93 for additional discussion and references.

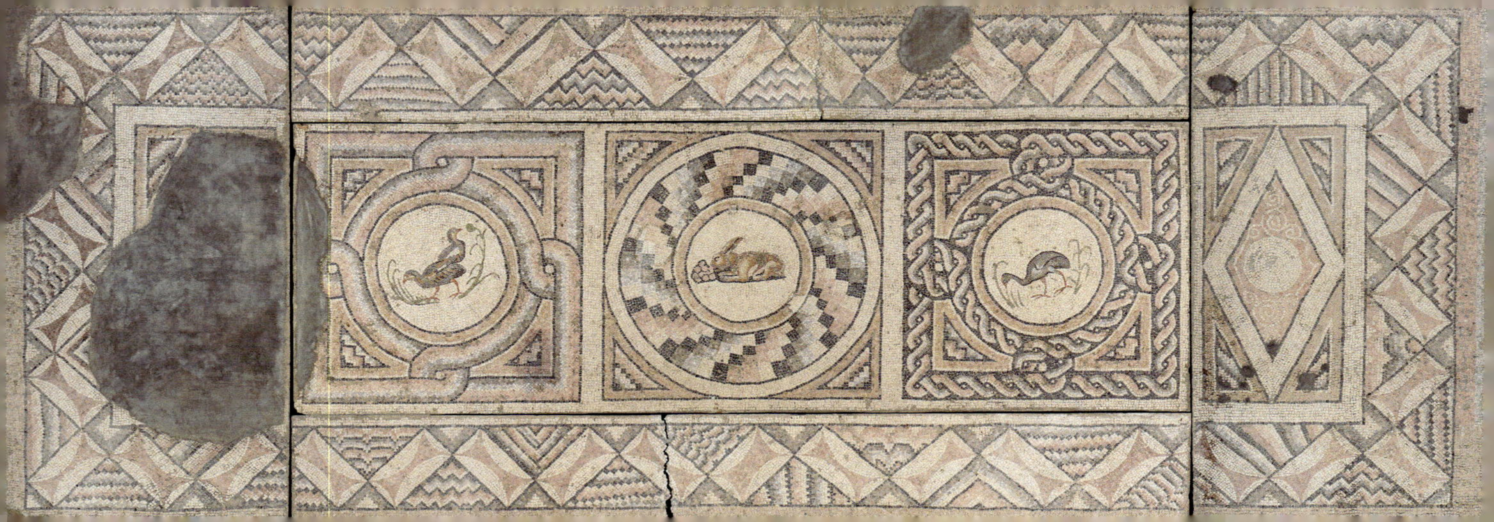

7. Mosaic Floor with Animals

Roman, from Antioch, Syria (present-day Antakya, Turkey), ca. AD 400
Stone tesserae, 257.2 cm × 682.3 cm
70.AH.96.1–.5

Provenance

This mosaic floor comes from a small bath building that was discovered by a local farmer near Toprak-en-Narlidja in Syria, located on the slopes of Mount Silpios, east of the ancient city of Antioch. A modern farm complex built over the site had used the earlier building material extensively. The ancient structure, known as the Bath of Apolausis, was uncovered in 1938 during a series of excavations made at Antioch and its vicinity.[1] The bath has been dated to the late fourth century BC.[2] Parts of the building dating to the mid-fifth century AD are thought to have been restorations or additions. The mosaic itself was exported from Syria soon after it was excavated and brought to the Worcester Art Museum, but it was never put on display. The J. Paul Getty Museum purchased the mosaic from the Worcester Art Museum in 1970.

Commentary

The dominant characteristic of the mosaic is its mesmerizing assortment of geometric patterns, executed in a muted color scheme consisting primarily of varying shades of gray, pink, ocher, black, and white. The only figural elements are the three square panels at the center of the mosaic, which depict animals in medallions enclosed by elaborately decorated borders. The main scene of a hare or rabbit eating a bunch of grapes is framed on either side by representations of birds pecking at foliage: two partridges on the left and a woodcock on the right. Continuous decorative bands—one a *guilloche*, the other a single cable—encircle the birds and merge with an exterior square border of the same pattern to create an enclosure for each of the figural compositions. A striking circular pattern composed of alternating rays of multicolored squares highlights the central medallion; this pattern, referred to as "rainbow style," produces a distinctive kaleidoscopic effect. A series of wavy lines fills the triangular spandrels in the four corners of each of the squares, matching similar patterns in the outer border of the mosaic. Rectangular end panels on the left and right sides of the mosaic contain lozenges (a decorative motif found primarily in coffered ceilings) that enclose a circle with a leafless rinceau. Although a large section of the left panel is missing, the right side is essentially complete.[3] A continuous border composed of rectangles and triangles, each decorated with a different geometric pattern, outlines the entire mosaic. The rectangles, placed diagonally in alternating directions, are marked by an inner hourglass shape. A variety of linear patterns, including zigzags and wavy lines running in different directions, decorate the triangles. These designs, which have abundant parallels in other mosaics at Antioch, are possibly intended to simulate the marble veining of contemporary floor pavements in *opus sectile*.[4]

This mosaic pavement originally covered the floor of a vestibule leading to the main room of the so-called Bath of Apolausis. The vestibule was entered through a portico along the building's southern side and gave access to the largest room of the bath, a central octagonal space with large niches that functioned both as the *frigidarium* and as the main hall of the building (see fig. 18). Bathing in the Roman world served an important social function. It was a necessary part of the daily routine, and many bath complexes were also cultural institutions that contained audience halls, libraries, and rooms devoted to cult activities. A distinguishing characteristic of baths in the Eastern Roman Empire in late antiquity was that the *frigidarium* transformed from a major hall with cold-water plunges to a spacious lounge-*apodyterium*, a hybrid room for gathering and entertainment.[5] The excavations of the area surrounding this bath indicate that it was a public facility, connected to a large country villa or a farm complex, one of several similar constructions in the vicinity.[6]

Although small in scale, the exceptional quality of the construction of the bath building and its lavish decoration, which included frescoes on the walls as well, attests to the wealth and taste of prosperous landowners in the countryside surrounding Antioch. The mosaic floors of the Bath of Apolausis illustrate themes that in late antiquity related to the beneficial effects of the baths, whose revitalizing waters could give pleasure and relieve pain.[7] The octagonal room and the adjoining apsidal room to the west were each decorated with an elaborate mosaic floor depicting a female bust. The figures are identified by inscriptions as the Greek personifications of Soteria (see fig. 19) (Salvation) and Apolausis (see fig. 20) (Enjoyment), qualities closely associated with the building.[8] Antioch's many sources of water—the Mediterranean Sea, the Orontes River, and especially the natural springs and pools at nearby Daphne, known for their healing properties—greatly enhanced the city's reputation for providing a luxurious quality of life. The leisure activities offered by public baths are praised by the fourth-century AD orator Libanius in his tribute to Antioch, which he describes as having the most beautiful and abundant waters of all cities, and therefore the finest baths.[9]

Comparanda

Representations of the central image on the Getty's mosaic floor, a crouching hare or rabbit feeding on fruit, appear as early as the first century AD in Roman wall paintings.[10] The diffusion of the motif is evident in mosaics found in a variety of contexts throughout the Roman Empire, dating from the third through the fifth century AD. In addition to the public bath building near Antioch, where the Getty mosaic was discovered, the image appears in mosaics from the basilicas at Aquileia in northern Italy and the hall of a public building at Sepphoris in Israel.[11] This theme was also popular in private settings—for example, an elaborate mosaic floor decorating the reception hall of a large villa at Lydda (present-day Lod) also in Israel. Mosaics from wealthy Roman villas at Corinium Dobunnorum (present-day Cirencester) in Britain and at Thysdrus (present-day El Djem) and Hadrumetum (present-day Sousse) in Tunisia depict similar scenes.[12] Hares with fruit frequently are found in vine-rinceau borders, as in the mosaic from the House of the Bird-Rinceau at Antioch, which includes birds, hares, and other animals within scrolls, between leaves and bunches of grapes.[13] The significance of the motif is not always clear, but it may relate to the cycle of life, as hares and rabbits are often associated with vitality and fertility.[14] This symbolic meaning, appropriate also in the decoration of the Bath of Apolausis and its reference to the good life, may have contributed to the continued use of the image in later Christian iconography, which centered on themes of eternal life.

Condition

The mosaic is mostly intact. A section (approximately 92 × 150 cm) on the left end of the mosaic is missing, however, and was reconstructed with concrete fill. The area of loss includes nearly all of one of the two rectangular panels with lozenge decorations.

Bibliography

Stillwell 1941, 19–23, fig. 21, 182–83, no. 123; Levi 1947, 304–5, plates 68b and 121a–b; Vermeule and Neuerburg 1973, 55, no. 116; Jones 1981, 20, M143, A–E; Balty 1995, 92–93; Cimok 2000, 238.

A.B.

NOTES

1. The campaigns were undertaken from 1932 to 1939. At the time, the Department of Syrian Antiquities permitted the sponsors of the excavations to export the mosaics. The excavation report was published in Stillwell 1941, 19–23, fig. 21, 182–83, no. 123.
2. Levi suggests a date of ca. AD 400, based on coins dating to the second half of the fourth century AD; see Levi 1947, 304, 626. Balty supports this on the basis of stylistic comparisons with other mosaic pavements of the region—particularly those from the Church of Khirbet Muqa in Apamea (AD 394/395); see Balty 1995, 92, and especially 89–93, for additional discussion and references.
3. Compare the image of the Getty mosaic in cat. 7 with the state of the mosaic in situ in fig. 17.
4. For additional examples from Antioch that may imitate this style, see Levi 1947, e.g., plates 73, 81, 109, 113. Balty also compares the patterns of additional mosaics in the Bath of Apolausis with pavements from other locations in Syria; see Balty 1995, 92–93. For a general catalogue of mosaic patterns at Antioch, see Campbell 1988, 85–100.
5. For more on baths and bathing in Roman Antioch, see Yegül 2000, 146–51.
6. The bath building formed the northern part of a courtyard, and a number of additional residential complexes were discovered to the east and the south; see Stillwell 1941, 19–23.
7. Personifications of abstract concepts, often associated with the function of the buildings they decorated, appear frequently in mosaics of the Eastern Roman Empire during this period. In another example from Antioch, a personification of Ananeosis (Renewal) may celebrate the restoration of the building in which it appeared; see Levi 1947, 320–21, plate 73.
8. The mosaic of Soteria is now in the Hatay Archaeological Museum in Antakya, Turkey (inv. no. 977, excavation no. 5287-M82A). The Apolausis mosaic is at Dumbarton Oaks Research Library and Collection (BZ.138.72). The use of Greek inscriptions to name figures is a common feature of mosaics. Personifications were ubiquitous in the fifth century AD, and the similarities between busts may have required labels to distinguish the various abstract concepts portrayed.
9. Libanius, *Oration*, 11.244–45.
10. A wall painting from Herculaneum showing a hare crouching beside figs is now in the Museo Archeologico Nazionale di Napoli (inv. no. 8630). This motif also appears in paintings in tombs—for example, a tomb discovered in Rome between the Appian Way and the Via Latina—see Reinach 1922, 364, 11.
11. For the basilicas at Aquileia (fourth century AD), see Cecchelli 1933, plate 17, 1 (North Basilica, *campata* V), and plate 37, 1 (South Basilica, *campata* III). For the third-century AD Roman public building at Sepphoris, see Talgam and Weiss 2004, 1–2, figs. 1, 2.
12. For the villa at Lod (fourth century AD), see Talgam and Weiss 2004, 12–13, fig. 13. For the mosaic from the villa at Corinium Dobunnorum (fourth century AD), found during the excavations at Beeches Road in 1971, see McWhirr 1986, 81, fig. 67 (now in the Corinium Museum, Cirencester). For the individual panel of a threshold mosaic at El Djem, see *Inv. Tunisie*, suppl., no. 71c, and for

the mosaic with dancing satyrs and nymphs at Sousse, see *Inv. Tunisie*, no. 155.

13. Levi dates this mosaic to the second quarter of the sixth century AD; see Levi 1947, plates 91–92. For additional examples dating from the Byzantine period, see Hachlili 2009, 154, figs. 7-2a–c.

14. Hares, prized hunting quarry well known for their fertility, were frequently depicted as an attribute of Aphrodite and served as a gift between lovers.

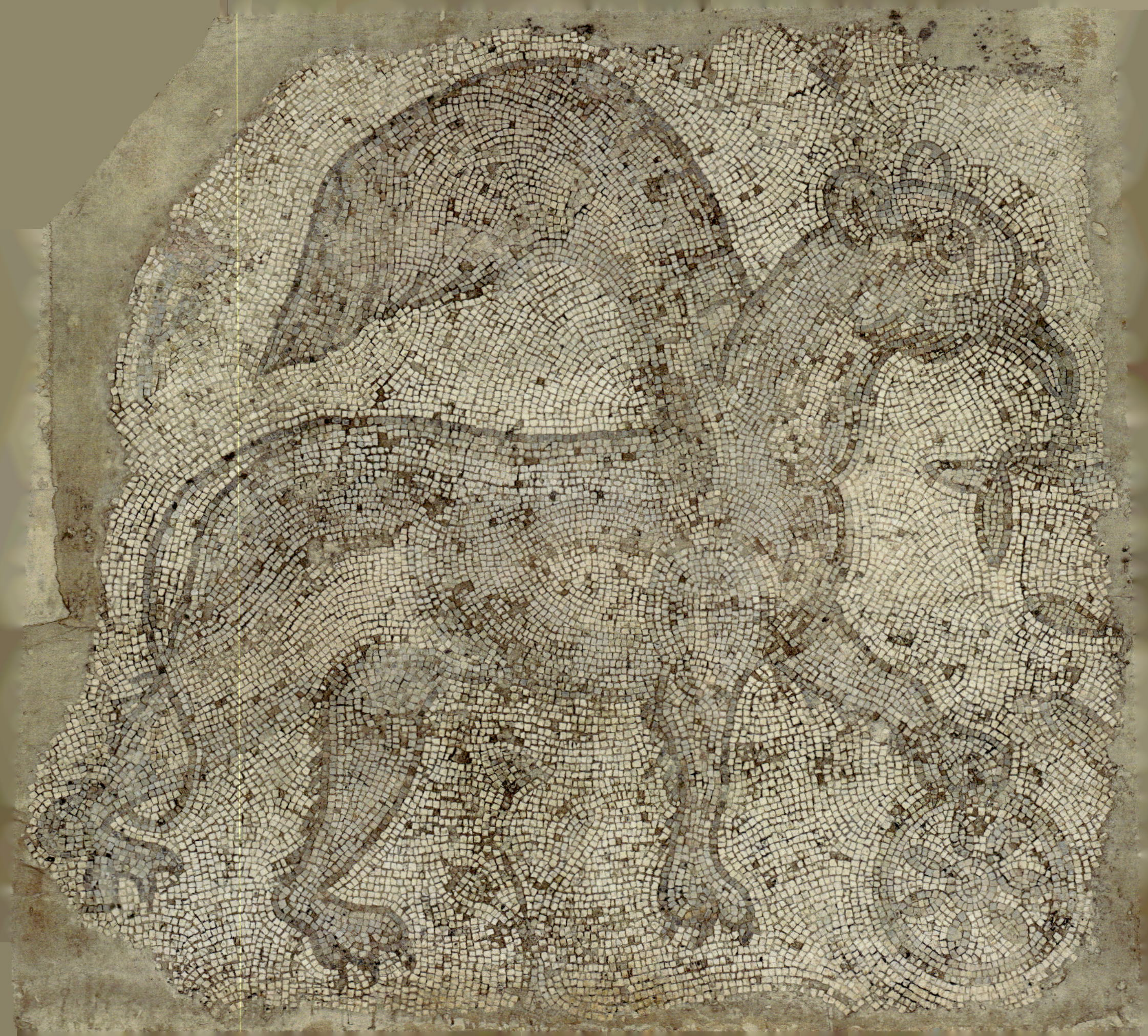

8. Mosaic Panel with Griffin

Roman, from Syria, possibly Emesa (present-day Homs), AD 400–600
Stone and glass tesserae, 160 cm × 173.4 cm
71.AH.113

Provenance

The origin of this mosaic is unknown, but it may have come from Emesa (present-day Homs), Syria, on the Orontes River, about 175 kilometers south of ancient Antioch.[1] The mosaic was purchased by the J. Paul Getty Museum in 1971.[2]

Commentary

A griffin, a mythological creature with the head of an eagle and the body of a winged lion, stands in profile with its raised left forepaw resting on top of a wheel with seven petal-shaped spokes.[3] The figure, depicted on a white background, is rendered primarily in neutral shades of gray and brown. The mosaic resembles a drawing, with a line of darker tesserae used to outline the griffin and define details, such as feathers at the tips of the wings and the underside of the head, as well as a fringe of fur along the leonine back legs.

The image of a griffin supporting one of its forepaws on a wheel appears in Roman art by the first century AD.[4] The wheel, a symbol of the cyclical movement of human fortune, and the winged griffin are both distinctive attributes of Nemesis, the goddess of vengeance, who is also often represented with wings.[5] In a first-century AD wall painting from the House of the Fabii at Pompeii, Apollo and two female figures are accompanied by a winged griffin with a wheel.[6] This motif also occurs on coins of Alexandria dating to the reign of the emperor Domitian (AD 81–96).[7] Scenes depicting Nemesis with a griffin are especially common during the second and third centuries AD and occur in many different media, including coins, gems, statues, and funerary and votive reliefs.[8] The particular image of a griffin resting its paw on a wheel, typically seated at the foot of Nemesis, is so pervasive that it eventually became a symbol for the goddess herself. For example, a limestone mold of the second to third centuries AD from Egypt, possibly from Alexandria, shows a griffin and a wheel with the Greek inscription *Nemesis*.[9]

Representations of the griffin with a wheel unaccompanied by Nemesis, as in the Getty mosaic, are particularly common in North Africa and the eastern periphery of the Roman Empire. The motif appears in the second and third centuries AD in Egyptian statuettes in faience and bronze; relief stelai from the amphitheater at Leptis Magna in present-day Libya; tomb paintings in Jordan; a votive marble statue from Erez, Israel, bearing a dedicatory inscription in Greek (dated AD 210–211); gems from Caesarea Maritima in Israel and Gadara in Jordan; and terracotta tesserae from Palmyra.[10] While the worship of Nemesis was widespread across the Roman Empire, it was particularly prevalent in Egypt, where she had a pre-Roman cult, and in Syria and the surrounding regions, where she was associated with several important local deities, including the classical goddesses Tyche (personification of fortune) and Nike (personification of victory) and the Arabic deities Allath (goddess of war) and Manawat (goddess of fate).[11]

Comparanda

Although it has been proposed that the Getty's griffin mosaic originally decorated a church floor, based on a supposed connection between this panel and a group of five others said to have come from a Syrian church, there is no archaeological or stylistic evidence that links these mosaics.[12] While griffins are not uncommon in later Christian mosaics, appearing in a number of Syrian church floors of the fifth and sixth centuries AD, such as those from the North Church at Huarte, the Getty panel is the only known instance in mosaic art in which the griffin appears with the Nemesis wheel.[13] Nemesis herself had disappeared from Roman art by this time, her function as a deity controlling human destiny and divine retribution made obsolete by the rise of Christianity. Perhaps the closest iconographical parallel to the Getty mosaic is a griffin decorating the floor of the peristyle of the Great Palace at Constantinople, which stands alone in the same pose, with a raised forepaw but without the wheel.[14]

Condition

Prior to entering the Getty collection, the tesserae were veneered, set in a concrete frame reinforced with iron laths, and coated with a microcrystalline wax.

Bibliography

Vermeule and Neuerburg 1973, 55–56, no. 117; *LIMC* s.v. "Nemesis (in Peripheria Orientali)," 772, no. 20.

A.B.

NOTES

1. It has been suggested that the mosaic came from an early Christian church; see Katherine Dunbabin's 1984 consultant report in the files of the antiquities department of the J. Paul Getty Museum; and Vermeule and Neuerburg 1973, 55–56, no. 117. Vermeule and Neuerburg proposed this origin because the mosaic was offered to the Getty Museum in 1971 as part of a group of five other animal panels said to have come from a church in Homs (see n. 12 below). They proposed a date of AD 450–462 based on inscribed panels apparently also related to this group, but no further documentation is provided for the inscriptions, and the whereabouts of the panels are currently unknown. However, there is no evidence linking the Getty's griffin mosaic to this group.
2. The mosaic was purchased from Peter Marks, New York.
3. Vermeule and Neuerburg interpreted this wheel as a circle containing a flower with seven petals; see Vermeule and Neuerburg 1973, no. 117.
4. For a possible Hellenistic connection, compare a stele from Apollonia in Thrace (present-day Bulgaria) that depicts a griffin with a raised paw between two enthroned figures; see Franke 1983, 58–59, fig. 113. Hornum, however, questions the early date; see Hornum 1993, 25.
5. Although representations of Nemesis accompanied by a griffin do not appear until the second century AD, the iconography of Nemesis with a wheel is established by the first century BC. For Nemesis and the griffin, see Hornum 1993, 24–32, 318–20, esp. 318, for examples of a griffin with a wheel. On the wheel as an attribute of Nemesis, see Hornum 1993, 25–28, 322–25. For additional discussion on Nemesis and the griffin, see Simon 1962, 770–78; and Flagge 1975, 12–14, 34–43.
6. The figures have been interpreted as Apollo, Venus, and Vesper, see *LIMC* s.v. "Apollo/Apollon," 421. For an alternate view of the figures as Apollo, Dionysos, and Venus, see Elia 1962, 119–20. The painting has been dated to the reign of the emperor Vespasian (AD 69–79).
7. Hornum 1993, plate 1.6; and Flagge 1975, 115.
8. For examples with references, see *LIMC* s.v. "Nemesis"; and Hornum 1993, 318.
9. This mold is now in the British Museum (1910, 0414.4); see Flagge 1975, 114–15, fig. 137; Hornum 1993, 318, plate 5; and *LIMC* s.v. "Nemesis," 754, no. 213, with additional references.
10. Egyptian faience statuette in the Brooklyn Museum (53.173): Flagge 1975, fig. 140. Bronze statuette in the Museum of Fine Arts, Boston (67.645): Comstock and Vermeule 1971, no. 173. Bronze statuette in Paris: Flagge 1975, fig. 141. Stelai from Leptis Magna: Hornum 1993, appendix 2, no. 159. Tomb paintings in Jordan: *LIMC* s.v. "Nemesis (in Peripheria Orientali)," 772, no. 19; and Barbet and Vibert-Guigue 1988, plates 105b, 122. Statue in Erez (now in the Israel Museum, Jerusalem [IAA 1957-866]): *LIMC* s.v. "Nemesis (in Peripheria Orientali)," 772, no. 23; and Leibovitch 1958, 141–48, plate 25. Jasper from Caesarea Maritima in Israel: Hamburger 1968, 32.95, plate 4.95. Carnelian from Gadara in Jordan: Henig and Whiting 1987, 37.396, fig. 396. Palmyrene tesserae: *LIMC* s.v. "Nemesis (in Peripheria Orientali)," 772, nos. 21–22; and Ingholt, Seyrig, and Starcky 1955, 28 (no. 194), 32 (no. 229), 44 (nos. 314–16) 46 (no. 331), 135, 1008.
11. An inscription dating to 110 BC from Memphis (Hornum 1993, appendix 2, no. 51) and a first-century BC papyrus (Papyrus Berlin 13954) (Hornum 1993, appendix 2, no. 60); both indicate a temple of Nemesis in Alexandria. Hornum also discusses the figure of Nemesis with the griffin holding a wheel in second-century AD statues from Cyrene and Side; see Hornum 1993, 19–24. On cults of the goddess in the East, see the discussion in *LIMC* s.v. "Nemesis (in Peripheria Orientali)," 772–73.
12. The mosaics were offered as a group, together with the Getty mosaic, by Peter Marks, New York. However, these panels are markedly different in style from the Getty mosaic. The Mosaic with a Lion is now in the Kimbell Art Museum, Fort Worth (AP 1972.17). The four additional panels are now in the Chazen Museum of Art at the University of Wisconsin–Madison: Mosaic of a Leopard Chasing a Gazelle (1972.17, Elvehjem Art Center 1973, fig. 8); Mosaic of a Dog Chasing a Rabbit (1972.18); Mosaic of an Amphora, Doves, and Peacocks (1972.19); and Mosaic of a Cock, Bird, Pheasant, Bull, and Deer in a Vine Scroll (1972.20, Elvehjem Art Center 1973, fig. 9).
13. Examples in Syrian church mosaics include scenes of Adam naming the animals and a griffin attacking a bull from the North Church at Huarte (Donceel-Voûte 1988, 105, fig. 71, and 114, fig. 80) and a mosaic from Apamea (Donceel-Voûte 1988, 207, figs. 187–88), all dated to the late fifth or early sixth century AD. An image of a griffin also appears earlier, in the early fourth-century AD mosaic of the Great Hunt of Piazza Armerina; see Gentili 1954, fig. 17.
14. See Brett 1947, 79, no. 50, plate 36; and Jobst and Vetters 1992, dated to the first half of the sixth century AD.

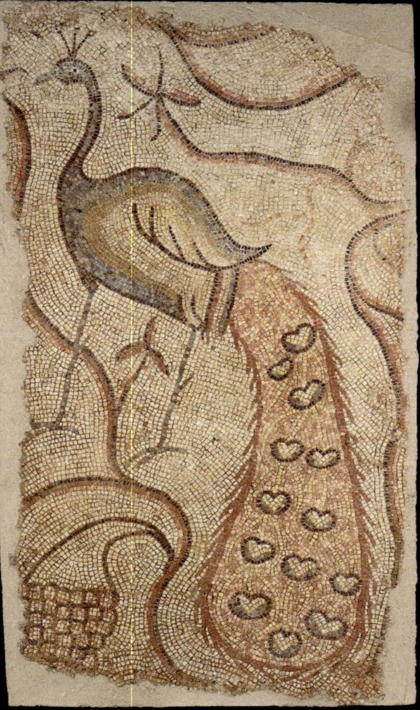
9

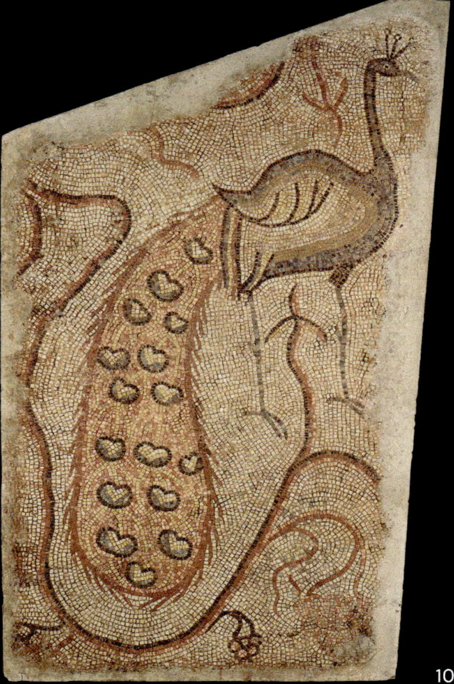
10

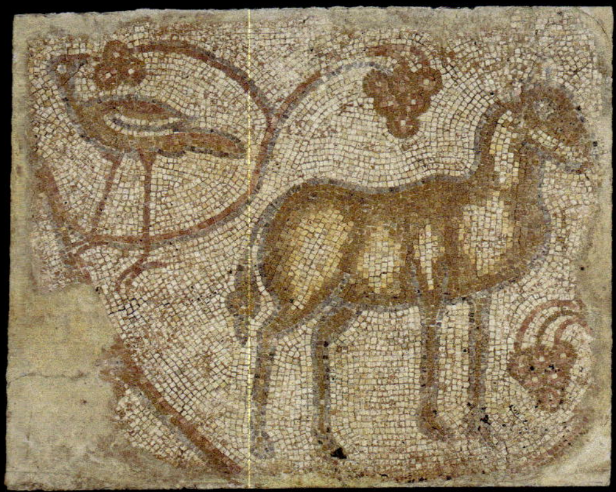
11

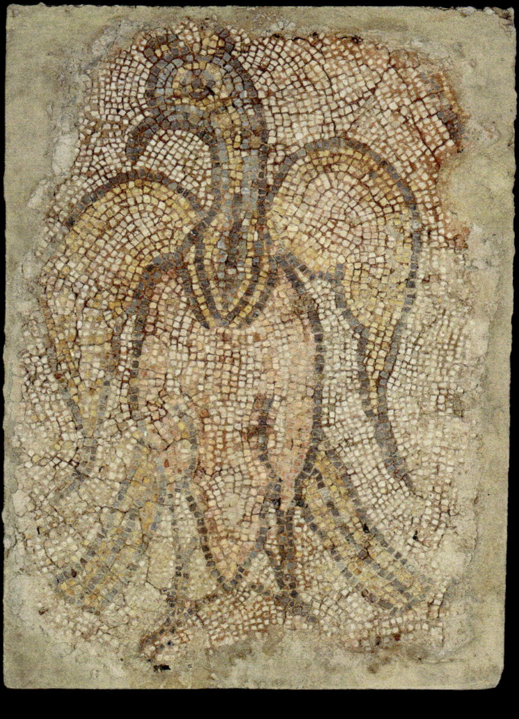
12

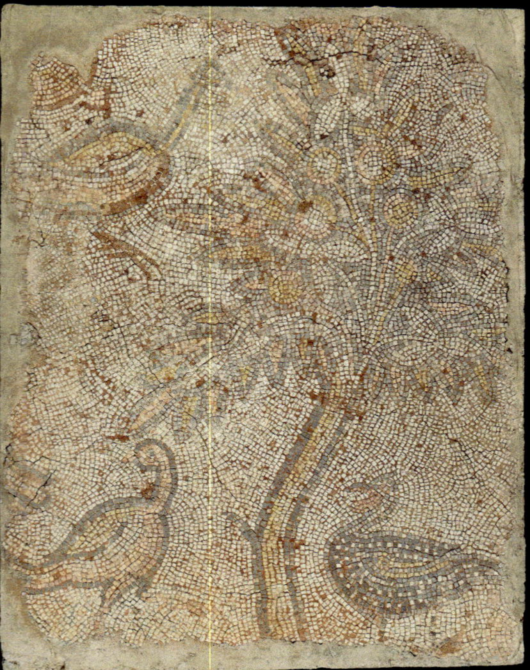
13

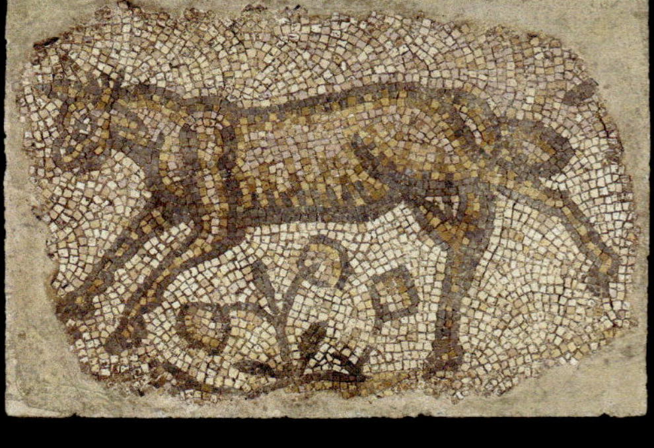
14

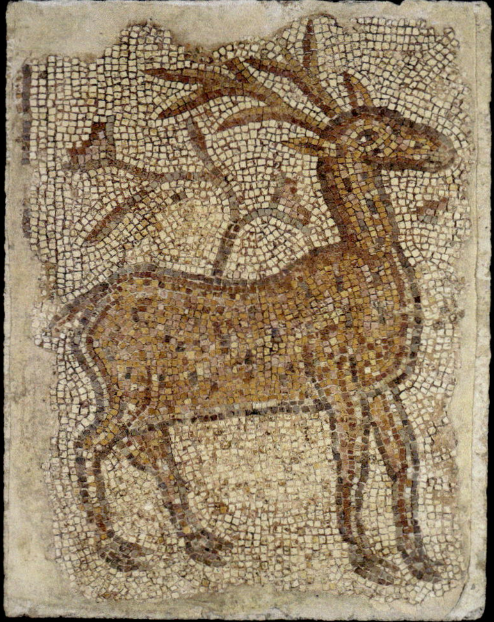

15

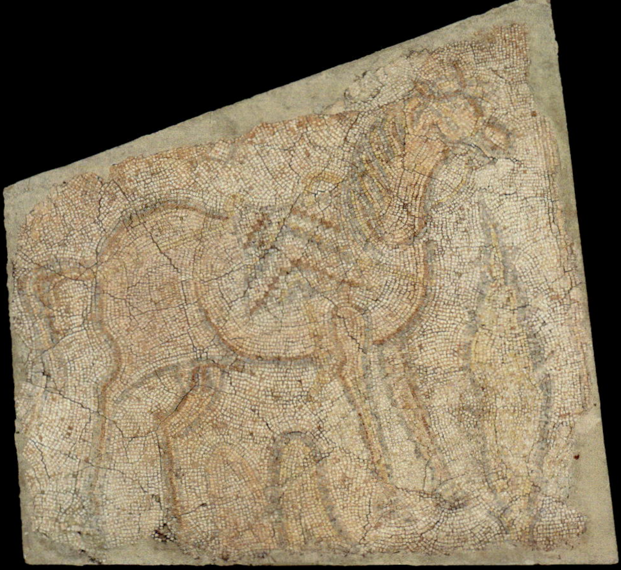

16

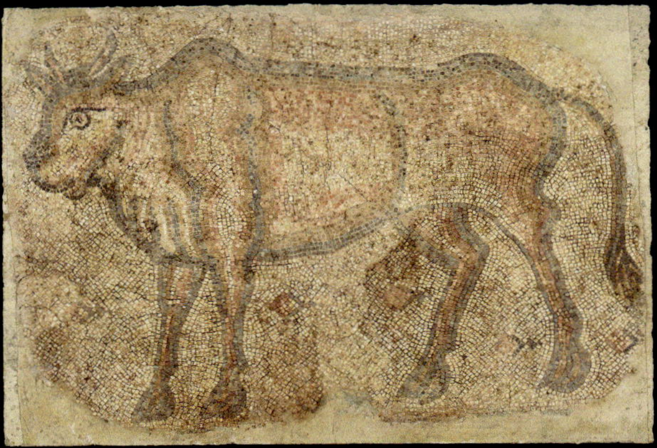

17

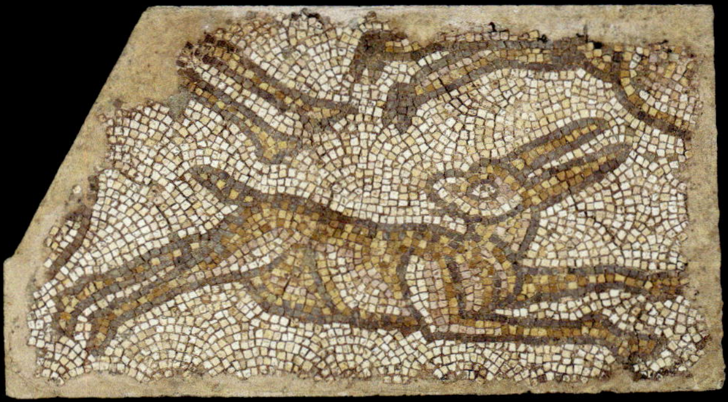

18

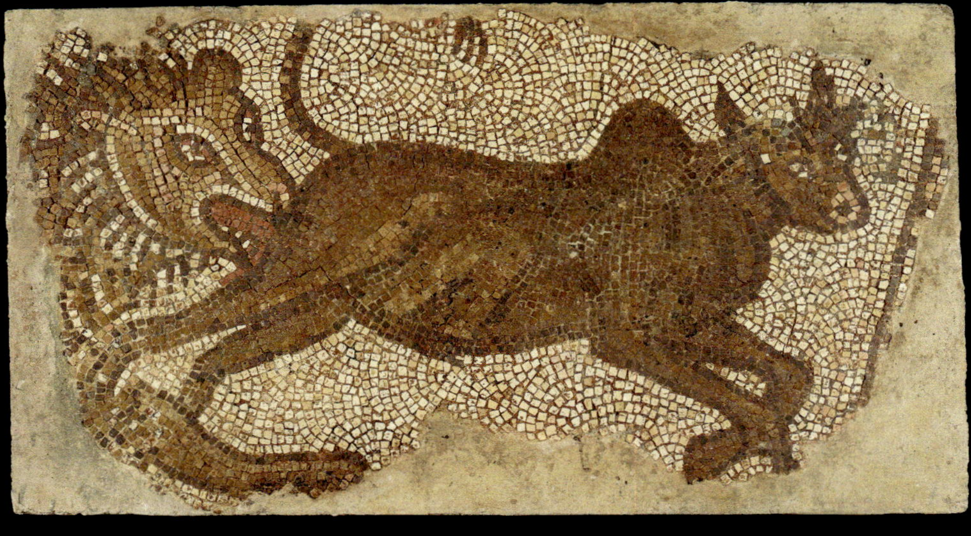

19

9–19. Mosaic Panels with Animals

9. Mosaic Panel with Peacock Facing Left
Roman, from Syria, possibly Emesa (present-day Homs), AD 400–600
Stone tesserae, 196.9 cm × 115.5 cm
75.AH.121
Gift of William Wahler

10. Mosaic Panel with Peacock Facing Right
Roman, from Syria, possibly Emesa (present-day Homs), AD 400–600
Stone tesserae, 210.2 cm × 129.5 cm
75.AH.122
Gift of William Wahler

11. Mosaic Panel with Donkey and Bird
Roman, from Syria, possibly Emesa (present-day Homs), AD 400–600
Stone tesserae, 91.4 cm × 106.7 cm
75.AH.119
Gift of William Wahler

12. Mosaic Panel with Eagle
Roman, from Syria, possibly Emesa (present-day Homs), AD 400–600
Stone tesserae, 81.3 cm × 57.2 cm
75.AH.125
Gift of William Wahler

13. Mosaic Panel with Three Birds and a Tree
Roman, from Syria, possibly Emesa (present-day Homs), AD 400–600
Stone tesserae, 125.1 cm × 97.8 cm
75.AH.123
Gift of William Wahler

14. Mosaic Panel of Bull and Flowers
Roman, from Syria, possibly Emesa (present-day Homs), AD 400–600
Stone tesserae, 75 cm × 109.2 cm
75.AH.124
Gift of William Wahler

15. Mosaic Panel with Stag
Roman, from Syria, possibly Emesa (present-day Homs), AD 400–600
Stone tesserae, 119.4 cm × 89 cm
75.AH.120
Gift of William Wahler

16. Mosaic Panel with Horse
Roman, from Syria, possibly Emesa (present-day Homs), AD 400–600
Stone tesserae, 147.3 cm × 146 cm
75.AH.116
Gift of William Wahler

17. Mosaic Panel with Bull
Roman, from Syria, possibly Emesa (present-day Homs), AD 400–600
Stone tesserae, 124.5 cm × 179.1 cm
75.AH.117
Gift of William Wahler

18. Mosaic Panel with Rabbit
Roman, from Syria, possibly Emesa (present-day Homs), AD 400–600
Stone tesserae, 68.6 cm × 114.3 cm
75.AH.118
Gift of William Wahler

19. Mosaic Panel with a Lion Chasing a Bull
Roman, from Syria, possibly Emesa (present-day Homs), AD 400–600
Stone tesserae, 81.3 cm × 149.9 cm
75.AH.115
Gift of Joel Malter

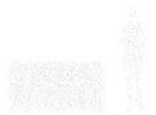

Provenance

The mosaics may have come from a church in Emesa (present-day Homs), Syria. However, there is no documentary evidence to confirm this provenance. The eleven panels were given to the J. Paul Getty Museum in 1975.[1]

Commentary

These eleven fragmentary mosaics represent various animals—bulls, a lion, a horse, a rabbit, a donkey, a stag, an eagle, peacocks, and other birds—and likely adorned the floors of one or more churches in Syria. Animals set among landscapes and vegetal motifs were typical of church mosaics in this region. Although all are very similar in style, the Getty mosaics feature a variety of compositions, including the so-called inhabited vine scroll, occasional landscape elements, such as trees and flowers, and both standing animals and active group scenes.

Vine scrolls decorate several of the Getty panels.[2] They are especially prominent on the two mosaics with peacocks (cats. 9, 10). Long vines with grape clusters wind across the surface and surround each bird. Peacocks frequently took a prominent position in church mosaics and were often paired in a symmetrical arrangement that distinguished them from other animals on the floor. These panels once flanked a tree or a wine vessel (traditionally, a *kantharos* or an *amphora*), which is now lost. Similar peacocks flank a *kantharos* in the sixth-century AD floor mosaic of a chapel at 'Ain el-Bad in Syria, now in the National Museum in Damascus, and in the sixth-century AD pavement of the "Armenian Chapel" near the Damascus Gate in Jerusalem.[3] The Mosaic Panel with Donkey and Bird (cat. 11) features a vine scroll bearing plump clusters of grapes. A bird perched on the vine turns its head to peck at one of the bunches of grapes. The Mosaic Panel with Eagle (cat. 12), which depicts the bird frontally, its wings outstretched in a heraldic pose, may have been framed by a vine scroll or a medallion, the remnants of which are visible around the borders of the panel.[4]

Other mosaic panels depict animals and birds in different landscape settings. The Mosaic Panel with Three Birds and Tree (cat. 13) shows two birds, one dark with white spots, the other light in color, sitting on the ground on either side of a tree, while a third bird perches above. The Mosaic Panel of Bull and Flowers (cat. 14), which portrays a running bull, and the Mosaic Panel with Stag (cat. 15) each contain a flowering bush. Similar bushes are present in a number of Syrian church mosaics, including the nave mosaics of the eighth-century Church of Saint George at Deir al-Adas.[5] On the Mosaic Panel with Horse (cat. 16), a horse wearing a rainbow-patterned saddle composed of triangles in alternating colors stands among several bushes or small trees. A similarly patterned horse's saddle is found in the mosaics of a church at Ma'arata.[6] On the Mosaic Panel with Bull (cat. 17), a muscular bull stands peacefully in a field of pink flowers, a feature also seen in the pavement of a fifth- or sixth-century AD church at Mezra'a el-'Ulia and in the fifth-century mosaics of the church at Temanaa, now in the Ma'arat an-Nu'man Museum.[7]

The Getty mosaic panels utilize various organizational schemes. While several depict individual animals standing alone, other panels show animals engaged in lively chase scenes. In the Mosaic Panel with Rabbit (cat. 18), a rabbit glances back as it leaps away from a predator, possibly a dog.[8] The legs of two additional running animals are preserved along the broken upper edge of the mosaic—on the left, the front hooves of one animal are visible, and on the right, the back legs of another are seen. In the Mosaic Panel with a Lion Chasing a Bull (cat. 19), a lion bites into the hindquarters of a bull as it tries to flee. The motif of a lion attacking a bull was common in the classical tradition, and it continued to be popular in Byzantine practice. The image is found, for example, in the aisle mosaics of the Church of Saint George at Houad (southeast of Apamea) in Syria, dated by inscription to AD 568, in which the lion bites into the bull, bringing forth large drops of blood.[9] The animals depicted on these two Getty panels are stylistically similar in their outlines and shading and can be compared to those depicted in the Mezra'a el-'Ulia pavement, which also features an animal chase.[10]

The range of different compositions represented in the Getty mosaics suggests that the panels came either from different churches or from several different spaces within a single church.[11] Churches in the region sometimes included vine scrolls and landscape scenes, like those on the Getty panels, in separate spaces of the same building. In the Church of 'Umm Hartaine in Syria, dated to ca. 500 AD, for instance, the apse mosaic is decorated with an inhabited scroll and the nave with animals chasing within a landscape setting.[12] The panels of a stag and of a lion chasing a bull (see cats. 15, 19) feature remnants of borders, suggesting that the mosaics may have come from panels placed in an aisle. In the aforementioned church at Houad, animals appear in bordered panels in the aisles and in a stacked arrangement in the nave.[13] The popular decorative scheme of two peacocks flanking a vessel (see cats. 9, 10) typically appeared in churches either at the entrance (west end) of the nave or in the sanctuary.[14] Other panels, such as the one with a horse (see cat. 16), may have been placed in an aisle or in the nave of a church, although their fragmentary state makes their original contexts difficult to determine. Rather than bordered panels, another possible layout is the free-form arrangement seen in the mosaics of a church near Hama (now in the Hama Archaeological Museum), which depicts animals walking or chasing each other around the nave.

In their Christian setting, the animals and other decorative elements would have alluded to Christian teachings. The vine scroll may have reminded viewers of Jesus's words in the Gospel according to John: "I am the true vine" (John 15:1). In early Christian art and literature, peacocks were considered among the most spectacular creatures on earth, and Augustine noted the belief that their flesh was incorruptible.[15] The eagle as a Christian symbol has been variously interpreted in connection with Christ and in its association with other birds, such as the Phoenix.[16] However, rather than conveying individual symbolic meanings, the majority of the animals represented in the Getty panels likely belonged to a larger scheme, in which the beasts represented the creatures in paradise or the variety of God's creations on earth. A similar selection of animals appears in several Syrian churches on mosaic floors that portray the scene of Adam naming the animals (Genesis 2:19–20).[17]

While the relationship between the Getty panels is uncertain, they do appear to have been produced either by the same group of mosaicists or by one or more closely related workshops.[18] The artisans employed a similar style and mannerisms, including heavy outlining of forms, rough modeling of body parts and musculature, and a color palette of reds, browns, beiges, and other earth tones. Although the execution of many of these panels does not show as much skill as that of the mosaics produced in metropolitan centers such as Antioch, the works do testify to the vibrant rural production of mosaics in the Levant in the fifth and sixth centuries AD.

Comparanda

In addition to the examples discussed above, stylistically similar panels are on display in the Ma'arat an-Nu'man Mosaics Museum, near Hama in Syria, as well as in a number of museums in the United States, including the Chazen Museum of Art at the University of Wisconsin–Madison, the Kimbell Art Museum in Fort Worth, and the Fordham Museum of Greek, Etruscan and Roman Art in New York.[19]

Condition

The panels are in varying states of preservation, with modern repairs of significant cracks and losses. All have been set into reinforced concrete frames, with the concrete mixture used to back and fill the panels, sometimes extending over the tesserae. Other fill materials used prior to the panels' arrival at the Getty include grout and a black (perhaps tar-based) caulk. The panels have been coated with a protective microcrystalline wax.

Bibliography

Unpublished.

S.L.

NOTES

1. The Mosaic with a Lion Chasing a Bull (cat. 19) was given to the Getty Museum by Joel Malter, Los Angeles, and the ten remaining panels were given to the museum by William Wahler, San Francisco, who purchased them from Mr. Malter.
2. On vine scrolls, see Talgam 2014, 86–96.
3. 'Ain el-Bad: Donceel-Voûte 1988, 16–19, fig. 1. Armenian Chapel in Jerusalem: Talgam 2014, 91–92, fig. 125. For another example, see the mosaic panel from a Syrian church in the Chazen Museum of Art (1972.19).
4. On depictions of eagles, often shown frontally with wings outstretched, see Hachlili 2009, 141.
5. Donceel-Voûte 1988, 49, fig. 23.
6. Donceel-Voûte 1988, 464, fig. 442b.
7. Jwejati 2014.
8. On the popular motif of a rabbit being chased by dogs, see Hachlili 2009, 157–58.
9. Donceel-Voûte 1988, 144, fig. 116.
10. Donceel-Voûte 1988, 179, fig. 150.
11. On inhabited scroll mosaics, which were especially popular in the sixth century AD, see Dauphin 1978, 1987; Hachlili 2009, 111–47; and Talgam 2014, 86–96.
12. Donceel-Voûte 1988, 192–201.
13. Donceel-Voûte 1988, 138–44. On the composition of animal and animal chase scenes in general, see Lavin 1963; Donceel-Voûte 1988, 476–78; and Hachlili 2009, 155–69.
14. See above, note 2.
15. Augustine, *De civitate Dei*, 21.4. On the Christian significance of peacocks, see Maguire 1987, 39–40.
16. On the eagle as the image of the bird freed from its cage by Christ, see Evans 1982, 219. Talgam reads the eagle in connection with the phoenix; see Talgam 2014, 201–2. For the eagle as a bird within a "catalogue" of birds; see Hachlili 2009, 141. The eagle has also been seen as a symbol of Christ the emperor, the congregation at communion, or the resurrection of Christ; see Maguire 1987, 65–66.
17. See Maguire 1987.
18. On mosaic workshops in the Levant, about which much remains unclear, see Hachlili 2009, 254–73; and Talgam 2014, 170–74. On mosaic workshops generally, see Donderer 1989.

19. In the Ma'arat an-Nu'man Mosaics Museum, for example, compare the nave pavement of the Houad church; see Donceel-Voûte 1988, 138–44. Mosaics with various animals are in the Chazen Museum (1972.17–20), formerly the Elvehjem Art Center. They are little documented, but see Elvehjem Art Center 1972–73, 42, ills. 8–9. A mosaic with a lion is now in the Kimbell Art Museum (AP 1972.17). Though the mosaics in the Fordham Museum are largely undocumented, their inscriptions were recently published; see Peppard 2014.

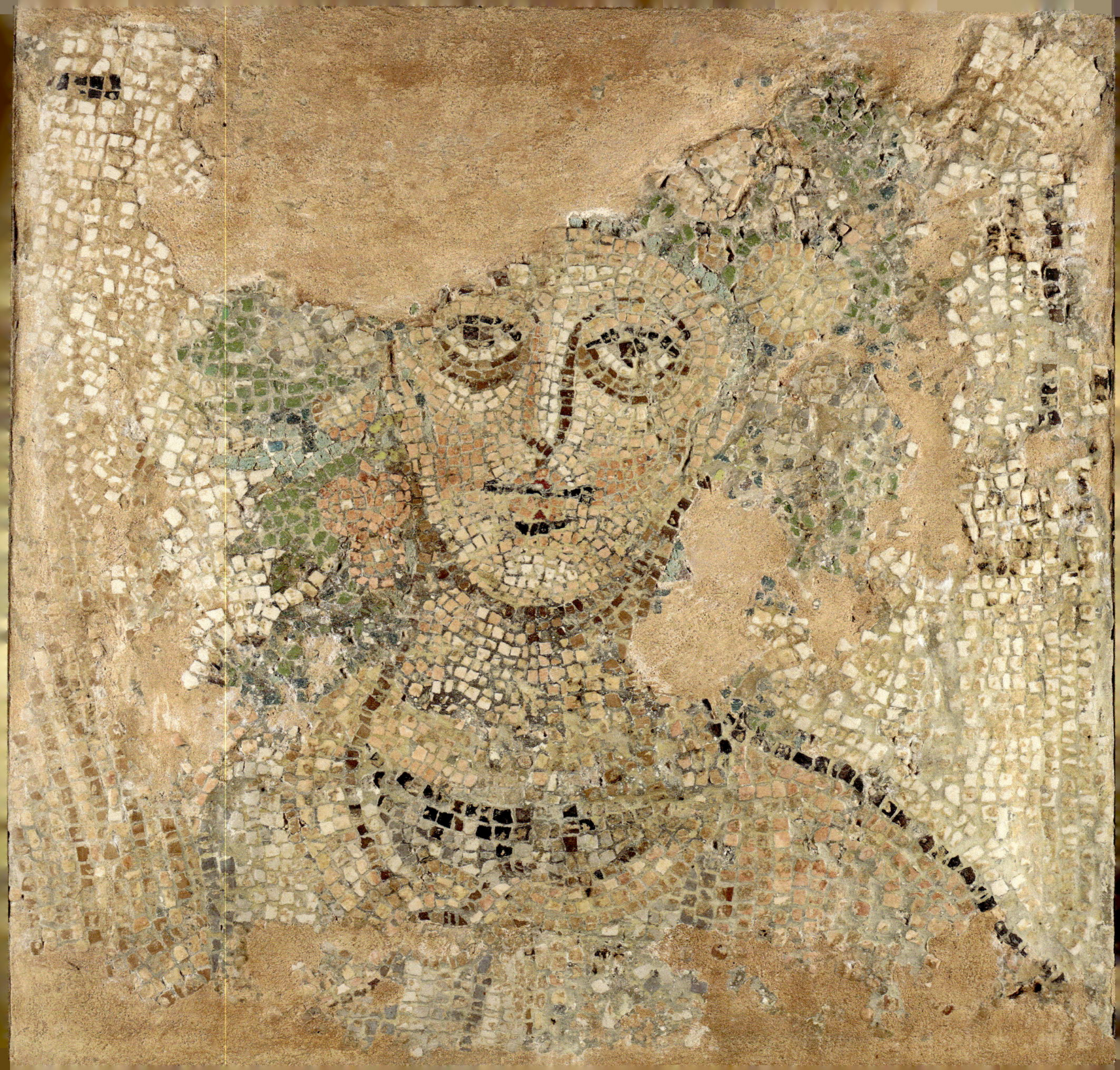

20. Mosaic Panel with Head of a Season

Roman, from Syria (or possibly Jordan), AD 400–600
Stone and glass tesserae, 73.7 cm × 74.3 cm
70.AH.95

Provenance

The findspot of this mosaic is unknown but is most likely Syria or possibly Jordan. The J. Paul Getty Museum purchased the mosaic in 1970.[1]

Commentary

A female bust crowned by a wreath of leaves, fruit, and flowers appears on this mosaic panel. Although previously identified as either Christ or Bacchus, the figure is more probably a personification of one of the Seasons, perhaps Methoporinē (Autumn in Greek).[2] Traces of the dark lines of a border run along the right and top left sides of the panel, suggesting that it was once part of a larger mosaic pavement that included personifications of Spring, Summer, and Winter. The features of the figure's face, especially the large, prominent eyes, straight nose, and heart-shaped lips, are characteristic of the regional style of the Levant, as seen in fifth- and sixth-century AD mosaics from sites such as the Nile Festival Building at Sepphoris in Israel and the Hippolytus Hall at Madaba in Jordan.[3] However, the range of colors represented in the Getty mosaic is unusual for mosaics of the region, as it includes more shades of blue and green made possible by the use of glass tesserae. While glass tesserae were more frequently seen in wall mosaics due to their relative expense and fragility, they were also used in floor mosaics to expand the mosaicists' palettes.[4]

The iconographical tradition of the Seasons is long and varied, and in the Roman period these figures could be depicted as either female or male.[5] By the fifth century AD, however, artists stopped producing images of male Seasons almost entirely, choosing instead to portray the figures exclusively as female.[6] Although of pagan origin, these personifications were judged acceptable for inclusion in both Christian and Jewish decorative programs in late antiquity, and they appear in funerary art (notably sarcophagi) and in church and synagogue floor mosaics.[7] As part of a larger program, this panel, from the floor of either a house or a Christian church, likely symbolized the order of the natural cycle of the year, presided over by God.[8]

Comparanda

Similar female busts of the Seasons appear in a number of mosaic pavements in Israel, Jordan, Lebanon, and Syria, typically on the floors of churches. At Deir es-Sleib in Syria, the four Seasons, each labeled with the corresponding name in Greek, appear in square panels surrounding a central roundel that decorates the pavement of the south sacristy of "Basilica A." Autumn wears a mantle tied in front in the same manner as that depicted on the Getty panel.[9] A mosaic representing the female personification of Ktisis (the act of donation), now in the Metropolitan Museum of Art in New York and likely from the same general region as the Getty mosaic, features a similar mantle.[10] Other extant mosaics that depict the Seasons include pavements at Caesarea Maritima, El-Maqerqesh, Madaba, and Petra.[11] While the Seasons are typically identifiable by their attributes—Autumn wearing her crown of leaves and fruit, Winter in her veil, Spring with her wreath of flowers, and Summer in her wreath of wheat—they are also usually labeled with their names in Greek, as at Deir es-Sleib. This kind of doubling of image and text served not only to identify the figures but also to exhibit the *paideia* (learnedness) of the patron.[12]

Condition

The mosaic is set in concrete and shows numerous signs of modern restoration, including the detachment and reintegration of a section on the bottom left, which has skewed the figure's right side at an angle to the rest of the body. The glass tesserae in the mosaic are extremely worn. Some original tesserae at the top of the panel are covered by the concrete fill.

Bibliography

Vermeule and Neuerburg 1973, 56, no. 118.

S.L.

NOTES

1. Purchased from Michel Dumez-Onof, London.
2. Vermeule and Neuerburg 1973, 56, no. 118. On the attributes of Autumn, see Parrish 1984, 38–40; and Hachlili 2009, 184–91. The personification of the Earth (Gē) is often affiliated, and sometimes conflated, with that of Autumn; see Talgam 2014, 350–51.
3. Sepphoris, Nile Festival Building: see Weiss and Talgam 2002; and Weiss 2009. Madaba, Hippolytus Hall: see Piccirillo 1993, 66.

4. While Vermeule and Neuerburg state that the inclusion of glass implies that the panel was made for a wall, the worn state of the glass tesserae, as well as the overall iconography, suggests that the panel was part of a floor mosaic; see Vermeule and Neuerburg 1973, 56. On the use of glass in floor mosaics, see Dunbabin 1999, 101–29.
5. Hanfmann 1951.
6. Hanfmann identifies at least twenty busts of the female Seasons but only six full-length male Seasons; see Hanfmann 1951, 264.
7. Hanfmann 1951, 262–67.
8. In Jewish contexts, the Seasons typically appear in triangular panels at the corner of circular depictions of the Zodiac; see Hachlili 2009, 45–48, 184–85. On late antique Christian interpretations of the Seasons, see Hanfmann 1951, especially 201, 205–6; Maguire 1987; and Talgam 2014, 191–94.
9. Donceel-Voûte 1988, 61–69, fig. 35.
10. Metropolitan Museum of Art 1998.69, 1999.99; see Evans, Holcomb, and Hallman 2001, 16–17.
11. See Hachlili 2009, 184–91. For examples in North African mosaics, see Parrish 1984, 204–6, no. 50, plate 69.
12. Leader-Newby 2005.

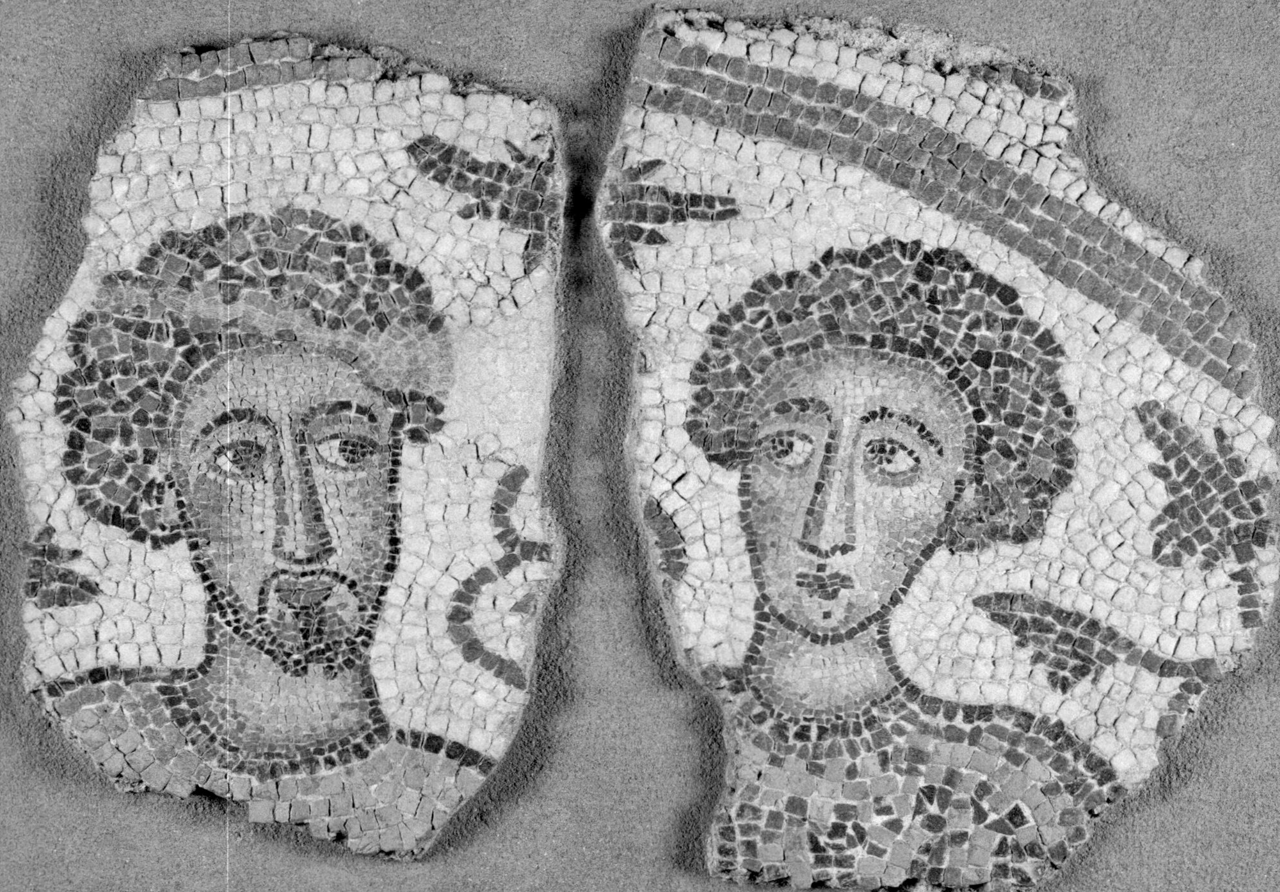

21. Mosaic Panel with Two Male Busts

Roman, from Syria (or possibly Jordan), AD 400–600
Stone tesserae, 48 cm × 68 cm
78.AH.399
Gift of Dr. Martin M. Orenstein

Provenance

The findspot of this mosaic is unknown but is most likely Syria or possibly Jordan. The mosaic was given to the J. Paul Getty Museum in 1978.[1]

Commentary

Two male busts appear in the middle of a vine scroll on this fragment of a mosaic pavement. Originally, this piece was part of a large floor mosaic, probably from a church, displaying the so-called inhabited scroll motif, a vine scroll that winds its way around the main field of the pavement.[2] Only a portion of the scroll is visible on this panel, but its presence suggests that the two men, one beardless and the other bearded, were participants in a series of scenes placed within circular medallions defined by the scroll. The men may have been depicted engaging in one of several typical rural activities, perhaps hunting, playing music, or harvesting grapes. A mosaic in the nave of the sixth-century AD Church of Saints Lot and Procopius at Khirbet Mukhayyat (Nebo) in Jordan shows villagers hunting lions and bears and participating in the grape harvest.[3] Such scenes of rural life, common in churches of the region in the later fifth and sixth centuries AD, would have evoked for contemporary audiences the activities and rhythms of their daily lives. The vine scroll may have also conveyed a religious meaning, alluding to Christ as the "true vine" (John 15:1).[4]

Comparanda

The quality of the mosaic is typical of rural workshops of the region, and the two faces are executed with much smaller tesserae than are found in the rest of the panel, a technique especially common in pavements from Jordan.[5] The panel is similar in style to that of many mosaics from the region, most notably in the narrow faces, heart-shaped lips, rosy cheeks, and schematized curly hair, characteristics that are shared with a Syrian mosaic depicting Adam and Eve that is now in the Cleveland Museum of Art.[6] Figures in other mosaics of the region, such as an image of a hunter in the church at Kissufim in Israel, have similar facial features.[7]

Condition

The mosaic is set in concrete and shows evidence of modern restoration, including areas painted to fill losses, such as the hair of the figure on the left.

Bibliography

Unpublished.

S.L.

NOTES

1. From 1975 to 1977, the mosaic was on loan to the Getty from Bruce McNall. It was purchased in 1977 by Dr. Martin M. Orenstein, who donated it to the J. Paul Getty Museum in 1978.
2. The "inhabited," or "peopled," scroll was especially popular in the region in the sixth century AD; see Hachlili 2009, 111–47; Talgam 2014, 86–95; and Dauphin 1978, 1987.
3. On these scenes of rural life, see Hachlili 2009, 149–78; on the church at Khirbet Mukhayyat, see Piccirillo 1993, 153, fig. 202.
4. See Maguire 1987, 9–10.
5. The use of small tesserae in this manner is seen in many churches in the region, including the church at Khirbet Mukhayyat, in the faces of the male figures; see Piccirillo 1993, 153.
6. Cleveland Museum of Art 1969.115; see Donceel-Voûte 1988, 487–89, 489 fig. 456.
7. See Hachlili 2009, plate 7.13a.

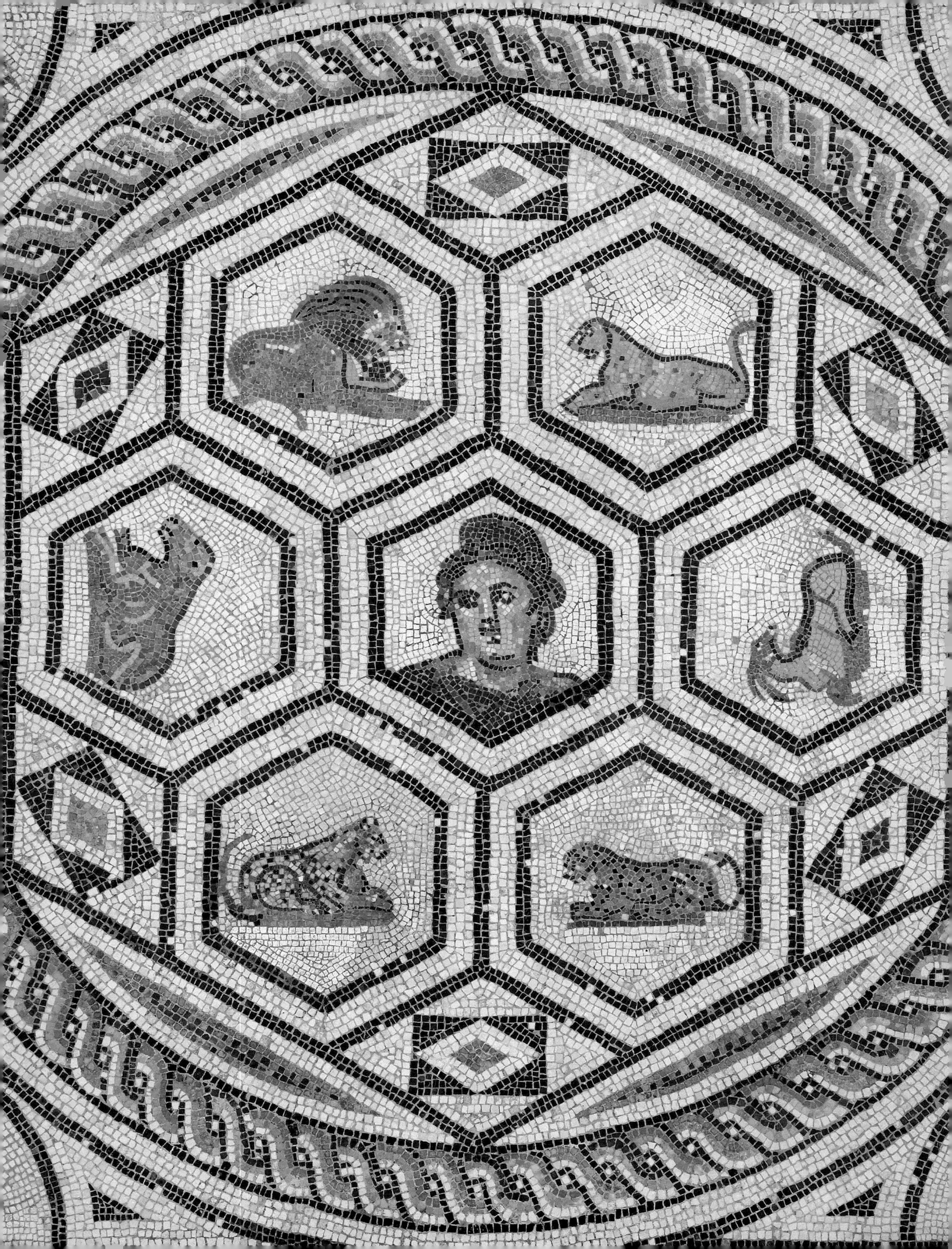

References Cited

Aisenberg 1994
Aisenberg, Nadya. *Ordinary Heroines: Transforming the Male Myth.* New York: Continuum, 1994.

Anderson 1985
Anderson, John Kinloch. *Hunting in the Ancient World.* Berkeley: University of California Press, 1985.

Armstrong 1986
Armstrong, Arthur H. *Classical Mediterranean Spirituality: Egyptian, Greek, Roman.* New York: Crossroad, 1986.

***ARV*[2]**
Beazley, John Davidson. *Attic Red-Figure Vase-Painters.* 2nd ed. Oxford: Clarendon Press, 1963.

Aurigemma 1961
Aurigemma, Salvatore. *Villa Adriana.* Rome: Istituto Poligrafico dello Stato, 1961.

Aurigemma 1974
Aurigemma, Salvatore. *The Baths of Diocletian and the Museo Nazionale Romano.* 7th ed. Rome: Istituto Poligrafico dello Stato, Libreria dello Stato, 1974.

Aymard 1937
Aymard, Jacques. "Quelques scènes de chasse sur une mosaïque de l'Antiquarium." *Mélanges d'archaéologie et d'histoire/École française de Rome* 54 (1937): 42–66.

Balil 1964
Balil, Alberto. "Notas iconograficas sobre algunos mosaicos de Sussa." *Les cahiers de Tunisie* 12, nos. 45–46 (1964): 7–10.

Ball 1984
Ball, David. "A Bear Hunt Mosaic." *J. Paul Getty Museum Journal* 12 (1984): 123–34.

Balty 1981
Balty, Janine. "Époque d'Hadrien: La mosaïque de Malibu (Californie)." In *La mosaïque antique au Proche-Orient, I: Des origines à la Tétrarchie*, 347–429. Vol. 2, part 12, 2, of *Aufstieg und Niedergang der Römischen Welt.* Berlin: W. de Gruyter, 1981.

Balty 1984
Balty, Janine. "Les mosaïques de Syrie au Ve siècle et leur répertoire." *Byzantion* 54 (1984): 437–68.

Balty 1995
Balty, Janine. *Mosaïques antiques du Proche-Orient: Chronologie, iconographie, interprétation.* Paris: Les Belles Lettres, 1995.

Barbet and Vibert-Guigue 1988
Barbet, Alix, and Claude Vibert-Guigue. *Les peintures des nécropoles romaines d'Abila et du nord de la Jordanie.* Vol. 2. Beirut: Institut Français d'Archéologie du Proche-Orient, 1988.

Becatti 1961
Becatti, Giovanni. *Scavi di Ostia.* Vol. 4, *Mosaici e pavimenti marmorei.* Rome: Libreria dello Stato, 1961.

Becker and Kondoleon 2005
Becker, Lawrence, and Christine Kondoleon. *The Arts of Antioch: Art Historical and Scientific Approaches to Roman Mosaics and a Catalogue of the Worcester Art Museum Antioch Collection.* Worcester, MA: Worcester Art Museum, 2005.

Bianchi Bandinelli 1955
Bianchi Bandinelli, Ranuccio. *Hellenistic-Byzantine Miniatures of the Iliad (Ilias Ambrosiana).* Olten, Switzerland: Urs Graf, 1955.

Bizot 1899
Bizot, E. "Découvertes de nouvelle Mosaïques a Sainte-Colombe." *Journal de Vienne et de l'Isère*, April 1, 1899, 1–2.

Bizot 1903
Bizot, E. "L'enlèvement d'Hylas: Mosaïque de Ste-Colombe." *Journal de Vienne et de l'Isère*, August 29, 1903, 1.

Blake 1936
Blake, Marion Elizabeth. "Roman Mosaics of the Second Century in Italy." *Memoirs of the American Academy in Rome* 13 (1936): 67–214.

Boriello and D'Ambrosio 1979
Boriello, M., and A. D'Ambrosio. *Baiae-Misenum.* Florence: L. S. Olschki, 1979.

Brett 1947
Brett, Gerard. *The Great Palace of the Byzantine Emperors, First Report.* London: Oxford University Press, 1947.

Bulletin de la Société nationale des antiquaires de France 1900
Bulletin de la Société nationale des antiquaires de France. "Séance du 18 Avril." *Bulletin de la Société nationale des antiquaires de France* (1900): 165–67.

Campbell 1988
Campbell, Sheila. *The Mosaics of Antioch.* Toronto: Pontifical Institute of Mediaeval Studies, 1988.

Cecchelli 1933
Cecchelli, Carlo. "Gli edifici e i mosaici paleocristiani nella zona della basilica di Aquileia." In *La Basilica di Aquileia*, 107–273. Bologna: N. Zanichelli, 1933.

Christou 1964
Christou, Chrysanthos. "Ἀνασκαφικαὶ εἰς περιοχὴν τῆς Σπάρτης." *Deltion* 19, no. 1 (1964): 135–38.

***CIL* IV**
della Corte, Matteo, and Pio Ciprotti, eds. *Corpus Inscriptionum Latinarum.* Vol. 4, *Inscriptiones parietariae Pompeianae, Herculanenses, Stabianae.* Suppl. 3.1, *Inscriptiones Pompeianae Herculanenses parietariae et vasorum fictilium.* Berolini: Walter de Gruyter, 1952.

Cima 1998
Cima, Maddalena. “Gli horti Liciniani: Una residenza imperiale della tarda antichità.” In *Horti Romani: Atti del convegno internazionale, Roma, 4–6 maggio 1995*, edited by Maddalena Cima and Eugenio La Rocca, 427–52. Rome: “L’Erma” di Bretschneider, 1998.

Cimok 2000
Cimok, Fatih, ed. *Antioch Mosaics*. Istanbul: A Turizm Yayınları, 2000.

Clarke 1979
Clarke, John. *Roman Black-and-White Figural Mosaics*. New York: New York University Press for the College Art Association of America, 1979.

Clarke 1991
Clarke, John. *The Houses of Roman Italy 100 BC–AD 250: Ritual, Space, and Decoration*. Berkeley: University of California Press, 1991.

Comptes-rendus de l’Academie des inscriptions et belles-lettres 1900
Comptes-rendus de l’Académie des inscriptions et belles-lettres. “Séance du 20 Avril.” *Comptes-rendus de l’Académie des inscriptions et belles-lettres* 44, no. 2 (1900): 222–28.

Comstock and Vermeule 1971
Comstock, Mary, and Cornelius Vermeule. *Greek, Etruscan, and Roman Bronzes in the Museum of Fine Arts, Boston*. Boston: Museum of Fine Arts, Boston, 1971.

Darmon 1990
Darmon, Jean-Pierre. “Les mosaïques du site des A.G.F.” In *Archéologie à Nîmes: Bilan de 40 années de recherches et découvertes, 1950–1990, ville de Nîmes, Musée archéologique*, 63–75. Nîmes: Musée Archéologique, 1990.

Darmon 1994
Darmon, Jean-Pierre. *Recueil général des mosaïques de la Gaule*. Vol. 2, *Province Lyonnaise*. Part 5, *Partie nord-ouest*. Paris: Centre National de la Recherche Scientifique, 1994.

Darmon 1995
Darmon, Jean-Pierre. “Les représentations mythologiques dans le décor domestique en narbonnaise à l’époque impériale.” *Revue archéologique de Picardie* 10 (1995): 59–74.

Dauphin 1978
Dauphin, Claudine. “Byzantine Pattern Books: A Re-examination of the Problem in the Light of the ‘Inhabited Scroll.’” *Art History* 1 (1978): 400–423.

Dauphin 1987
Dauphin, Claudine. “The Development of the Inhabited Scroll in Architectural Sculpture and Mosaic Art from Late Imperial Times to the Seventh Century A.D.” *Levant* 19, no. 1 (1987): 183–213.

de Villefosse 1899
de Villefosse, Antoine-Marie Héron. “Mosaïques trouvées a Sainte-Colombe-lès-Vienne.” *Bulletin archéologique du Comité des travaux historiques et scientifiques* 17 (1899): 102–3.

de Villefosse 1903
de Villefosse, Antoine-Marie Héron. “Rapport,” *Bulletin archéologique du Comité des travaux historiques et scientifiques* (1903): 109–19.

Donceel-Voûte 1988
Donceel-Voûte, Pauline. *Les pavements des églises byzantines de Syrie et du Liban: Décor, archéologie et liturgie*. Louvain-la-Neuve: Département d’Archéologie et d’Histoire de l’Art, 1988.

Donceel-Voûte 1994
Donceel-Voûte, Pauline. “Le Ve siècle dans les mosaïques syriennes.” In vol. 4 of *La mosaïque gréco-romaine*, edited by Jean-Pierre Darmon and Alain Rebourg, 205–10. Paris: Association Internationale pour l’Étude de la Mosaïque Antique, 1994.

Donderer 1989
Donderer, Michael. *Die Mosaizisten der Antike und ihre wirtschaftliche und soziale Stellung: Eine Quellenstudie*. Erlangen: Universitätsbund, 1989.

Ducrey, Metzger, and Reber 1993
Ducrey, Pierre, Ingrid R. Metzger, and Karl Reber. *Le quartier de la Maison aux mosaïques*. Lausanne: Éditions Payot, 1993.

Dunbabin 1978
Dunbabin, Katherine. *Mosaics of Roman North Africa: Studies in Iconography and Patronage*. Oxford: Clarendon Press; New York: Oxford University Press, 1978.

Dunbabin 1994
Dunbabin, Katherine. “Early Pavement Types in the West and the Invention of Tessellation.” In *Fifth International Colloquium on Ancient Mosaics, Held at Bath, England, on September 5–12, 1987*, edited by Peter Johnson, Roger Ling, and David Smith, 26–40. Ann Arbor, MI: Association for the Study and Preservation of Roman Mosaics, 1994.

Dunbabin 1999
Dunbabin, Katherine. *Mosaics of the Greek and Roman World*. Cambridge: Cambridge University Press, 1999.

Egea 1988
Egea, Michel. “Entente et Dares sauves de la terre.” *Le Provencal*, February 4, 1988.

Elia 1962
Elia, Olga. “Lo stibadio dionisiaco in pitture pompeiane.” *Mitteilungen des deutschen archäologischen Instituts. Römische Abteilung* 69 (1962): 118–27.

Elvehjem Art Center 1972–73
Elvehjem Art Center. “Part Three. Donors and Acquisitions.” *Elvehjem Art Center Bulletin* (1972–73): 37–47.

Espérandieu 1913
Espérandieu, Émile. *Recueil général des bas-reliefs, statues et bustes de la Gaule romaine*. Vol. 5, *Belgique*. Paris: Imprimerie Nationale, 1913.

Espérandieu 1935
Espérandieu, Émile. *Les mosaïques romaines de Nîmes*. Nîmes: Éditions Mëridionales, 1935.

Evans 1982
Evans, Helen C. “Nonclassical Sources for the Armenian Mosaic near the Damascus Gate in Jerusalem.” In *East of Byzantium: Syria and Armenia in the Formative Period*, edited by Nina G. Garsoïan, Thomas F. Mathews, and Robert W. Thomson, 217–22. Washington, DC: Dumbarton Oaks, Center for Byzantine Studies, 1982.

Evans, Holcomb, and Hallman 2001
Evans, Helen C., Melanie Holcomb, and Robert Hallman. "The Arts of Byzantium." *Metropolitan Museum of Art Bulletin* 58, no. 4 (2001): 16–17.

Fabia 1923
Fabia, Philippe. *Mosaïques romaines de museés de Lyon*. Lyon: Museés de Lyon, 1923.

Famà 1997
Famà, Maria Luisa. "Il mosaic a ciottoli di Mozia dopo il restauro." In *Atti del IV colloquio dell'Associazione italiana per lo studio e la conservazione del mosaic Palermo, 9–13 dicembre 1996*, edited by Rosa Maria Bonacasa Carra and Federico Guidobaldi, 147–58. Ravenna: Edizioni del Griasole, 1997.

Ferraro 1982
Ferraro, Salvatore. *La presenza di Virgilio nei graffiti pompeiani*. Naples: Loffredo, 1982.

Flagge 1975
Flagge, Ingeborg. *Untersuchungen zur Bedeutung des Greifen*. Sankt Augustin, Germany: Hans Richarz, 1975.

Foucher 1960a
Foucher, Louis. *Découvertes archéologiques à Thysdrus en 1960*. Tunisia: Imprimerie La Rapide, 1960.

Foucher 1960b
Foucher, Louis. *Inventaire des mosaïques, feuille no 57 de l'Atlas archéologique: Sousse*. Tunisia: Institut National d'Archéologie et Arts.

Foucher 1961
Foucher, Louis. *Découvertes archéologiques à Thysdrus en 1961*. Tunisia: Imprimerie du Secréteriat d'État aux Affairs Culturelles et à l'Information, 1961.

Foucher 1963
Foucher, Louis. *La maison de la procession dionysiaque à El Jem*. Paris: Presses Universitaires de France, 1963.

Fowler 1902
Fowler, Harold. 1902. "Archaeological News: Villelaure." *American Journal of Archaeology* 6 (1902): 370.

Franke 1983
Franke, P. R. "Albaniem im Altertum." *Antike Welt: Zeitschrift für Archäologie und Kulturgeschichte* 14 (1983): 11–64.

Fredericksen 1975
Fredericksen, Burton B., ed. *The J. Paul Getty Museum: Greek and Roman Antiquities, Western European Paintings, French Decorative Arts of the Eighteenth Century*. London: Weidenfeld and Nicolson, 1975.

Fredericksen, Frel, and Wilson 1980
Fredericksen, Burton B., Jiří Frel, and Gillian Wilson, eds. *The J. Paul Getty Museum: Guidebook*. 5th ed. Malibu: J. Paul Getty Museum, 1980.

Gabrici 1901
Gabrici, E. "Baia—Mosaico figurato scoperto presso le così dette Stufe di Nerone." *Notizie degli Scavi* (1901): 297–98.

Gaday et al. 2006
Gaday, Robert, et al. *La Tuilière à Villelaure (Vaucluse), rapport final d'opération diagnostic*. Nîmes: Institut National de Recherches Archéologiques Préventives, 2006.

Gauckler 1897
Gauckler, Paul. "Le domaine des Laberii à Uthina." *Monuments et memoires: Fondation Eugene Piot* 3 (1897): 177–226.

Gauckler 1901
Gauckler, Paul. "Note sur quelques mosaïques romaines de Provence." *Bulletin archéologique du Comité des travaux historiques et scientifiques* (1901): 336–46.

Gauckler 1910
Gauckler, Paul. *Inventaire des mosaïques de la Gaule ed de l'Afrique*. Vol. 2, *Afrique Proconsulaire (Tunisie)*. Paris: Ernest Leroux, 1910.

Gentili 1954
Gentili, Gino Vinicio. *La Villa imperiale di Piazza Armerina*. 2nd ed. Rome: Istituto Poligrafico dello Stato, Libreria dello Stato, 1954.

Gentili 1959
Gentili, Gino Vinicio. *La Villa Erculia di Piazza Armerina: I mosaici figurati*. Edited by Annibale Belli. Rome: Edizioni Mediterranee, 1959.

Getty 1965
Getty, J. Paul. *The Joys of Collecting*. New York: Hawthorn Books, 1965.

Getty and Le Vane 1955
Getty, J. Paul, and Ethel Le Vane. *Collector's Choice: The Chronicle of an Artistic Odyssey through Europe*. London: W. H. Allen, 1955.

Geymonat 1987
Geymonat, Mario, ed. *Eneide con episodi significativi di Iliade e Odissea*. Bologna: N. Zanichelli, 1987.

Hachlili 2009
Hachlili, Rachel. *Ancient Mosaic Pavements: Themes, Issues, and Trends*. Boston: Brill, 2009.

Hamburger 1968
Hamburger, Anit. *Gems from Caesarea Maritima*. Jerusalem: Department of Antiquities and Museums in the Ministry of Education and Culture; Jerusalem: Department of Archaeology, Hebrew University, 1968.

Hammer Galleries 1941
Hammer Galleries. *Select Art Objects from the William Randolph Hearst Collection*. New York: Hammer Galleries, 1941.

Hanfmann 1951
Hanfmann, George M. A. *The Season Sarcophagus in Dumbarton Oaks*. Cambridge, MA: Harvard University Press, 1951.

Hartmann 1930
Hartmann, Albert, ed. "Federzeichnung auf einem Münchener Papyrus." In *Festschrift für Georg Leidinger zum 60. Geburtstag am 30 Dezember 1930*, 103–8. Munich: H. Schmidt, 1930.

Heap 1883
Heap, G. H. "On an Antique Roman Mosaic from Carthage, Now in the United States National Museum." *Proceedings of the United States National Museum* 6 (1883): 415–17.

Helbig 1963
Helbig, Wolfgang. *Die Päpstlichen Sammlungen im Vatikan und Lateran.* Vol. 1, *Führer durch die öffentlichen Sammlungen klassicher Altertümer in Rom.* Tübingen: Ernest Wasmuth, 1963.

Henig and Whiting 1987
Henig, Martin, and Mary Whiting. *Engraved Gems from Gadara in Jordan: The Sa'd Collection of Intaglios and Cameos.* Oxford: Oxford University Committee for Archaeology; Oxford: Oxbow Books, 1987.

Hessenbruch McKeon 1983
Hessenbruch McKeon, Carolyn. *Iconology of the Gorgon Medusa in Roman Mosaic.* PhD diss., University of Michigan, 1983.

Hornum 1993
Hornum, Michael B. *Nemesis, the Roman State and the Games.* New York: E. J. Brill, 1993.

***IG* VII**
Dittenberger, Guilelmus, ed. *Inscriptiones Graecae.* Vol. 7, *Inscriptiones Megaridis et Boeotiae.* Berlin: Walter de Gruyter, 1978.

Ingholt, Seyrig, and Starcky 1955
Ingholt, Harald, Henri Seyrig, and Jean Starcky. *Recueil des tessères de Palmyre.* Paris: Paul Geuthner, 1955.

Jacquème 1922
Jacquème, C. *Histoire de Cadenet.* Part 1. Marseilles: Imprimerie de la Sté du "Petit Marseillais," 1922.

Jesnick 1997
Jesnick, Ilona Julia. *The Image of Orpheus in Roman Mosaic: An Exploration of the Figure of Orpheus in Graeco-Roman Art and Culture with Special Reference to Its Expression in the Medium of Mosaic in Late Antiquity.* Oxford: Archaeopress, 1997.

Jobst and Vetters 1992
Jobst, Werner, and Hermann Vetters. *Mosaikenforschung im Kaiserpalast von Konstantinopel: Vorbericht über das Forschungs- und Restaurierungsprojekt am Palastmosaik in den Jahren 1983–1988.* Vienna: Verlag der Österreichischen Akademie der Wissenschaften, 1992.

Jones 1981
Jones, Frances F. "Appendix: Mosaics from Antioch in Other American Institutions." *Record of the Art Museum, Princeton University* 40, no. 2 (1981): 16–20.

J. Paul Getty Museum 2010
J. Paul Getty Museum. *The J. Paul Getty Museum Handbook of the Antiquities Collection.* 2nd ed. Los Angeles: J. Paul Getty Museum, 2010

Jwejati 2014
Jwejati, Rafah. "Contribution de la mosaïque syrienne à l'iconographie chrétienne." In *The Levant: Crossroads of Late Antiquity: History, Religion and Archaeology,* edited by Ellen Bradshaw Aitken and John M. Fossey, 401–22. Leiden: Brill, 2014.

Kitzinger 1965
Kitzinger, Ernst. "Stylistic Developments in Pavement Mosaics in the Greek East from the Age of Constantine to the Age of Justinian." In *La mosaïque gréco-romaine. Paris, 29 août – 3 septembre 1963,* 1:341–52. Paris: Éditions du Centre National de la Recherche Scientifique, 1965.

Kolling 1982
Kolling, Alfons. *Der römische Palast von Nennig.* Munich: Zurich: Schnell & Steiner, 1982.

Kondoleon 2000
Kondoleon, Christine. "Mosaics of Antioch." In *Antioch: The Lost Ancient City,* edited by Christine Kondoleon, 63–77. Princeton: Princeton University Press, 2000.

König 2005
König, Jason. *Athletics and Literature in the Roman Empire.* Cambridge: Cambridge University Press, 2005.

Labande 1903
Labande, L.-H. "Les mosaïques romaines de Villelaure." *Bulletin archéologique du Comité des travaux historiques et scientifiques* (1903): 1–13.

Labande and de Villefosse 1900
Labande, L.-H., and Héron de Villefosse. "Rapport." *Bulletin archéologique du Comité des travaux historiques et scientifiques* (1900): 113–14.

***La dépêche de Lyon* 1913**
La dépêche de Lyon, August 2, 1913.

Lafaye 1901
Lafaye, Georges. "Rapport." *Bulletin de la Société nationale des antiquaires de France* (1901): 118–22.

Lafaye 1909
Lafaye, Georges. *Inventaire des mosaïques de la Gaule et de l'Afrique.* Vol. 1, *Narbonne et Aquitaine.* Paris: Ernest Leroux, 1909.

Lancha 1977
Lancha, Janine. *Mosaïques géométriques: Les ateliers de Vienne (Isère): Leurs modèles et leur originalité dans l'Empire romain.* Rome: "L'Erma" di Bretschneider, 1977.

Lancha 1981
Lancha, Janine. *Recueil général des mosaïques de la Gaule.* Vol. 2, *Province de Narbonnaise.* Part 2, *Vienne.* Paris: Éditions du Centre National de la Recherche Scientifique, 1981.

Lancha 1990
Lancha, Janine. *Les mosaïques de Vienne.* Lyon: Presses Universitaires de Lyon, 1990.

Lancha 1997
Lancha, Janine. *Mosaïque et culture dans l'Occident romain Ier–IV^e^s.* Rome: "L'Erma" di Bretschneider, 1997.

Langsner 1958
Langsner, Jules. "'An Astonishing Renascence ...' Mosaic: Why the Revival." *Los Angeles Times,* March 30, 1958.

Lapatin 2014
Lapatin, Kenneth. "The Getty Bear Hunt Mosaic: Past, Present, and Future." Abstract of a paper presented at a colloquium organized by the North American branch of the L'Association Internationale pour l'Étude de la Mosaïque Antique, Malibu, California, October 16–18, 2014, http://www.aiema-namerica.org/LAPATIN-MALIBU%20ABSTRACT.pdf.

Lassus 1934
Lassus, J. "La mosaïque de Yakto." *Antioch-on-the-Orontes* 1 (1934): 114–56.

Lavagne 1977
Lavagne, Henri. "Une mosaïque gallo-romaine au musée J. Paul Getty." *J. Paul Getty Museum Journal* 5 (1977): 177–82.

Lavagne 1978
Lavagne, Henri. "Le combat de Darès et Entelle (Enéide, V. v. 362–482) sur une mosaïque de Villelaure (Vaucluse)." In *Présence de Virgile: Actes du colloque des 9, 11 et 12 décembre 1976: Paris E.N.S., Tours*, edited by Raymond Chevallier, 133–46. Paris: Les Belles Lettres, 1978.

Lavagne 1979
Lavagne, Henri. *Recueil général des mosaïques de la Gaule*. Vol. 3, *Province de Narbonnaise*. Part 1, *Partie central*. Paris: Éditions du Centre National de la Recherche Scientifique, 1979.

Lavagne 1990
Lavagne, Henri. "La Maison I: Les Mosaïques." In *Les Fouilles de l'Enclos des Chartreux: De l'antiquité aux XVIIe siècle*, edited by Géraldine Bérand, Brigitte de Luca, and Corinne Landuré, 20–23. Aix-en-Provence: Conseil Général des Bouches-du-Rhône, 1990.

Lavagne 1994
Lavagne, Henri. "Le pavement de la rue des Magnans à Aix-en-Provence et la naissance des trames à décor multiple dans la mosaïque gallo-romaine." *Gallia* 51 (1994): 202–15.

Lavagne 2000
Lavagne, Henri. *Recueil général des mosaïques de la Gaule*. Vol. 3, *Province de Narbonnaise*. Part 3, *Partie sud-est*. Paris: Éditions du Centre National de la Recherche Scientifique, 2000.

Lavagne 2014
Lavagne, Henri. "Virgile à la table des notables." In *Aix en archéologie: 25 ans de découverts*, edited by Núra Nin, 196–98. Bruxelles: Snoeck, 2014.

Lavin 1963
Lavin, Irving. "The Hunting Mosaics of Antioch and Their Sources: A Study of Compositional Principles in the Development of Early Mediaeval Style." *Dumbarton Oaks Papers* 17 (1963): 179–286.

Leader-Newby 2005
Leader-Newby, Ruth. "Personifications and *Paideia* in Late Antique Mosaics from the Greek East." In *Personification in the Greek World: From Antiquity to Byzantium*, edited by Emma Stafford and Judith Herrin, 231–46. Aldershot, England: Ashgate, 2005.

Leader-Newby 2014
Leader-Newby, Ruth. "Heroes, Lions, and Vandals: Four Late Roman Missoria." In *The Berthouville Silver Treasure and Roman Luxury*, edited by Kenneth Lapatin, 89–105. Los Angeles: Getty Publications, 2014.

***Le bien public* 1961**
Le bien public, April 19, 1961.

Lefebvre 1981
Lefebvre, L. "Les sculptures gallo-romaines découvertes dans le sous-sol de la Breck à Arlon." *Bulletin de l'Institut archéologique du Luxembourg Arlon* 57, nos. 1–2 (1981): 34–54.

Lefebvre 1987
Lefebvre, Louis. "Un scène tirée de l'œuvre de Virgile est représentée sur le mausolée de Vervicius." *Bulletin de l'Institut archéologique du Luxembourg Arlon* 63, no. 1/2 (1987): 14–17.

Leibovitch 1958
Leibovitch, J. "Le Griffon d'Erez et le sens mythologique de Némésis." *Israel Exploration Journal* 8, no. 3 (1958): 141–48.

***Le Moniteur Viennois* 1903**
"Les Mosaïques de Saint-Romain-en-Gal." *Le Moniteur Viennois*, July 29, 1913, 2.

***Le Progrès* 1913**
"Ils en veulent en Angeleterre." *Le Progrès*, July 25, 1913.

Levi 1947
Levi, Doro. *Antioch Mosaic Pavements*. Princeton: Princeton University Press, 1947.

***LIMC* 1981–2009**
Lexicon Iconographicum Mythologiae Classicae. Zürich: Artemis Verlag, 1981–2009.

Ling 1998
Ling, Roger. *Ancient Mosaics*. Princeton: Princeton University Press, 1998.

***London News* 1966**
"Pieces for Collectors." *Illustrated London News*, November 26, 1966.

***Lyon républicain* 1913**
Lyon républicain, July 26, 1913.

Maguire 1987
Maguire, Henry. *Earth and Ocean: The Terrestrial World in Early Byzantine Art*. University Park: Pennsylvania State University Press, 1987.

Markoulaki 1987
Markoulaki, Stavroula. "Καστέλλι Κισάμου." *Deltion* 42, no. 2 (1987): 558–66.

McWhirr 1986
McWhirr, Alan. *Houses in Roman Cirencester*. Cirencester, England: Cirencester Excavation Committee, Corinium Museum, 1986.

Merlin 1915
Merlin, Alfred. *Inventaire des mosaïques de la Gaule et de l'Afrique*. Vol. 2, suppl., *Afrique Proconsulaire (Tunisie)*. Paris: Ernest Leroux, 1915.

Michon 1923a
Michon, Étienne. *Bulletin de la Société nationale des antiquaires de France* (1923): 109.

Michon 1923b
Michon, Étienne. Letter (unpublished), February 1923, in Archives du Musée du Louvre, Antiquités Grecques et Romaines, no. A 5, 28 février 1923.

Michon 1923c
Michon, Étienne. "Rapport." *Bulletin archéologique du Comité des travaux historiques et scientifiques* (1923): LX.

Milnor 2014
Milnor, Kristina. *Graffiti and the Literary Landscape in Roman Pompeii.* Oxford: Oxford University Press, 2014.

Minchev 2002
Minchev, Alexander. "The 'House of Antiope,' a Late Roman Residential House with Mosaics at Marcianopolis (Devnya, Bulgaria)." In *The Roman and Late Roman City: The International Conference, Veliko Turnovo, 26–30 July 2000*, edited by L. Ruseva-Slokoska, Rumen Teofilov Ivanov, and Ventsislav Dinchev, 245–54. Sofia: Professor Marin Drinov Academic Publishing House, 2002.

Ministère de la culture et communication 1981
Ministère de la culture et communication, Commission régionale de Provence-Alpes Côte d'Azur. *Inventaire général des monuments et des richesses artistiques de la France; Vaucluse: Cantons Cadenet et Pertuis: Pays d'Aigues.* Paris: Imprimerie Nationale, 1981.

Moretti and Sgubini Moretti 1977
Moretti, Mario, and Anna Maria Sgubini Moretti. *La Villa dei Volusii a Lucus Feroniae.* Rome: Autostrade, 1977.

Neuerburg 1975
Neuerburg, Norman. "Mosaic of a Lion Attacking a Horse." *J. Paul Getty Museum Journal* 2 (1975): 51.

Newby 2005
Newby, Zahra. *Greek Athletics in the Roman World: Victory and Virtue.* New York: Oxford University Press, 2005.

Özüdoğru and Dökü 2010
Özüdoğru, Şükrü, and F. Eray Dökü. "Kibyra 2009." *Anmed: Anadolu Akdenizi Arkeoloji Haberleri* 8 (2010): 38–45.

Özüdoğru and Dökü 2012
Özüdoğru, Şükrü, and F. Eray Dökü. "Kibyra 2011." *Anmed: Anadolu Akdenizi Arkeoloji Haberleri* 10 (2012): 46–52.

Pace 1955
Pace, Biagio. *I Mosaici di Piazza Armerina.* Rome: G. Casini, 1955.

Pagano 1983–84
Pagano, M. "Il lago Lucrino: Ricerche storiche e archeologiche." In Vols. 7, 8, *Puteoli, studi di storia antica.* Pozzuoli, Italy: Azienda Autonoma di Soggiorno, Cura e Turismo, 1983–84.

Panayotopoulou 1998
Panayotopoulou, Anastasia. "Roman Mosaics from Sparta." In *Sparta in Laconia: Proceedings of the 19th British Museum Classical Colloquium, Held with the British School at Athens and King's and University Colleges, London 6–8 December 1995*, edited by William G. Cavanaugh and S. E. C. Walker, 112–18. London: British School at Athens, 1998.

Paribeni 1913
Paribeni, Roberto. "Nuovi monumenti del Museo Nazionale Romano." *Bollettino d'arte* 7, no. 5 (1913): 157–72.

Parke-Bernet Galleries 1949
Parke-Bernet Galleries. *The Notable Art Collection Belonging to the Estate of the Late Joseph Brummer.* Part 1. New York: Parke-Bernet Galleries, 1949.

Parlasca 1959
Parlasca, Klaus. *Die römischen Mosaiken in Deutschland.* Berlin: De Gruyter, 1959.

Parrish 1984
Parrish, David. *Seasons Mosaics of Roman North Africa.* Rome: Giorgio Bretschneider, 1984.

Parrish 1987
Parrish, David. "A Mosaic of a Lion Attacking an Onager." *Karthago* 21 (1987): 113–34.

Pasqui 1911
Pasqui, Angelo. "Roma: Nuove scoperte di antichità nella città e nel suburbia." *Notizie degli Scavi* 8 (1911): 338–39.

Paton 2000
Paton, S. "The Villa Dionysos at Knossos: A *Domus* and Its Décor." In *Πεπραγμενα Η' Διεθνούς Κρητολογικού Συνεδρίου, Ηρακλείου 9–14 σεπτεμβρίου 1996* A2, 553–62. Herakleion: Εταιρία Κρητικών Ιστορικών Μελετών, 2000.

Peppard 2014
Peppard, Michael. "Mosaics from a Church in the Diocese of Apamea, Syria (463 CE)." *Zeitschrift für Papyrologie und Epigraphik* 190 (2014): 168–72.

Philadelpheus 1894
Philadelpheus, Alexandros. "Το Γοργόνειον εν τῷ εκ Πειραιώς Ψηφηιδωτῷ." *Archaiologikē ephēmeris* 3 (1894): 99–112.

Phillips 1960
Phillips, Kyle. "Subject and Technique in Hellenistic-Roman Mosaics: A Ganymede Mosaic from Sicily." *Art Bulletin* 42, no. 4 (1960): 243–62.

Piccirillo 1993
Piccirillo, Michele. *The Mosaics of Jordan.* Edited by Patricia M. Bikai and Thomas A. Dailey. Amman, Jordan: American Center of Oriental Research, 1993.

Pisapia 2014
Pisapia, M. S. "Uno sfortunata mosaico con scene di caccia da Baia." In *Atti del XIX Colloquio dell' Associazione italiana per lo studio e la conservazione del mosaico con il patrocinio del Ministero per i beni e le attività culturali (Isernia, 13–16 marzo 2013)*, edited by Claudia Angelelli, 109–19. Tivoli, Italy: Scripta Manent, 2014.

Poinssot and Lantier 1923
Poinssot, L., and R. Lantier. "Mosaïque de la chasse au sanglier à Carthage." *Bulletin de la Société nationale des antiquaires de France* 1923, 154–58.

Poliakoff 1985
Poliakoff, Michael B. "Entellus and Amycus: Virgil, *Aen.* 5. 326–484." *Illinois Classical Studies* 10, no. 2 (1985): 227–31.

Poliakoff 1987
Poliakoff, Michael B. *Combat Sports in the Ancient World: Competition, Violence, and Culture*. New Haven: Yale University Press, 1987.

Rebetez 1992
Rebetez, Serge. "Zwei figürlich verzierte Mosaiken und ein Lararium aus Vallon (Schweiz)." *Antike Welt: Zeitschrift für Archäologie und Kulturgeschichte* 23, no. 1 (1992): 15–29.

Reinach 1912
Reinach, Salomon. *Répertoire de reliefs grecs et romains*. Vol. 2, *Afrique, Îles Brittaniques*. Paris: Leroux, 1912.

Reinach 1922
Reinach, Salomon. *Répertoire de peintures grecques et romaines*. Paris: Leroux, 1922.

Rey 1979
Rey, Jean-Claude. *Histoire de Villelaure*. Villelaure: Credel, 1979.

Salzmann 1982
Salzmann, Dieter. *Untersuchungen zu den antiken Kieselmosaiken: Von den Anfängen bis zum Beginn der Tesseratechnik*. Berlin: Gebr. Mann, 1982.

Sautel 1939
Sautel, Joseph. *Carte et texte complet du département du Vaucluse*. Paris: Ernest Leroux, 1939.

Sauze and Muret 2008
Sauze, Élisabeth, and Jean-Pierre Muret. *Villelaure et ses châteaux*. Lauris, France: Association "Pour le Luberon," 2008.

Scheibelreiter-Gail 2011
Scheibelreiter-Gail, Veronika. *Die Mosaiken Westkleinasiens: Tessellate des 2. Jahrhunderts v. Chr. bis Anfang des 7. Jahrhunderts n. Chr.* Vienna: Österreichisches Archäologisches Institut, 2011.

Schwarzmaier, Scholl, and Maischberger 2012
Schwarzmaier, Agnes, Andreas Scholl, and Martin Maischberger, eds. *Die Antikensammlung: Altes Museum, Neues Museum, Pergamonmuseum*. Darmstadt: Philipp von Zabern, 2012.

Seignoret 1987
Seignoret, Chantal. "Le chasseur de mosaïques." *Vaucluse Matin*, September 26, 1987.

Shear 1930
Shear, Theodore Leslie. *The Roman Villa*. Cambridge, MA: Harvard University Press, 1930.

Simon 1962
Simon, Erika. "Zur Bedeutung des Greifen in der Kunst der Kaiserzeit." *Latomus* 21, no. 4 (1962): 749–80.

Slim 1987
Slim, Hédi. "Orpheus Charming the Animals." In *Carthage: A Mosaic of Ancient Tunisia*, edited by Aïcha Ben Abed Ben Khader and David Soren, 210–11. New York: American Museum of Natural History, 1987.

Sotheby Parke Bernet 1973
Sotheby Parke Bernet. *Egyptian, Western Asiatic, Byzantine, Islamic, Greek, Roman, Etruscan Antiquities, May 4, 1973*. New York: Sotheby Parke Bernet, 1973.

Spinazzola 1928
Spinazzola, Vittorio. *Le arti decorative in Pompei e nel Museo nazionale di Napoli*. Milan: Bestetti e Tumminelli, 1928.

Spink & Son, Ltd. 1966
Spink & Son, Ltd. *Octagon* 3, no. 4. London: Spink & Son, 1966.

Stern 1955
Stern, Henri. "La Mosaïque d'Orphée de Blanzy-lès-Fismes (Aisne)." *Gallia* 13 (1955): 41–77.

Stern 1971
Stern, Henri. "Mosaïques de la région de Vienne (Isère)." *Gallia* 29 (1971): 123–49.

Stillwell 1941
Stillwell, Richard, ed. *Antioch-on-the-Orontes*. Vol. 3, *The Excavations 1937–1939*. Princeton: Princeton University Press, 1941.

Strocka 1991
Strocka, Volker Michael. *Casa del Labirinto (VI 11, 8–10)*. Munich: Hirmer, 1991.

Talgam 2014
Talgam, Rina. *Mosaics of Faith: Floors of Pagans, Jews, Samaritans, Christians, and Muslims in the Holy Land*. University Park: Pennsylvania State University Press, 2014.

Talgam and Weiss 2004
Talgam, Rina, and Zeev Weiss. *The Mosaics of the House of Dionysos at Sepphoris*. Jerusalem: Institute of Archaeology, Hebrew University, 2004.

Tallah 2004
Tallah, Linda. *Le Luberon et pays d'Apt*. Paris: Académie des Inscriptions et Belles-Lettres, 2004.

Toynbee 1973
Toynbee, Jocelyn M. C. *Animals in Roman Life and Art*. Ithaca, NY: Cornell University Press, 1973.

Trilling 1989
Trilling, James. "The Soul of the Empire: Style and Meaning in the Mosaic Pavement of the Byzantine Imperial Palace in Constantinople." *Dumbarton Oaks Papers* 43 (1989): 27–72.

Tusa 1997
Tusa, Vincenzo. "I mosaic di Mozia." In *Atti del IV colloquio dell'Associazione italiana per lo studio e la conservazione del mosaic, Palermo, 9–13 dicembre 1996*, edited by Rosa Maria Bonacasa Carra and Federico Guidobaldi, 137–46. Ravenna: Edizioni del Griasole, 1997.

Vassy and Guy 1913
Vassy, Albert, and Claudius Guy. "Mise à jour et description de trois mosaïques romaines à Saint-Romain-en-Gal (Rhône) près Vienne (Isère)." *Association française pour l'avancement des sciences* (1913): 628–32.

Vermeule and Neuerburg 1973
Vermeule, Cornelius, and Norman Neuerburg. *Catalogue of the Ancient Art in the J. Paul Getty Museum: The Larger Statuary, Wall Paintings, and Mosaics*. Malibu: J. Paul Getty Museum, 1973.

von Boeselager 1983
von Boeselager, Dela. *Antike Mosaiken in Sizilien: Hellenismus und römische Kaiserzeit*, vol. 3. Rome: Giorgio Bretschneider, 1983.

von Gonzenbach 1975
von Gonzenbach, Victorine. "Ein neues Briseismosaik." In *La mosaïque gréco-romaine II: IIe Colloque international pour l'étude de la mosaïque antique, Vienne, 30 août–4 septembre 1971*, edited by Henri Stern and Marcel Le Glay, 401–8. Paris: A. & J. Picard, 1975.

Weiss 2009
Weiss, Zeev. "The Mosaics of the Nile Festival Building at Sepphoris and the Legacy of the Antiochene Tradition." In *Between Judaism and Christianity: Art Historical Essays in Honor of Elisheva (Elisabeth) Revel-Neher*, edited by Katrin Kogman-Appel and Mati Meyer, 7–24. Leiden: Brill, 2009.

Weiss and Talgam 2002
Weiss, Zeev, and Rina Talgam. "The Nile Festival Building and Its Mosaics: Mythological Representation in Early Byzantine Sepphoris." In *Late-Antique Petra, Nile Festival Building at Sepphoris, Deir Qal'a Monastery, Khirbet Qana Village and Pilgrim Site, 'Ain-'Arrub Hiding Complex, and Other Studies*, 55–61. Vol. 3, *The Roman and Byzantine Near East*, edited by John H. Humphrey. Portsmouth, RI: Journal of Roman Archaeology, 2002.

Werner 1994
Werner, Klaus. *Mosaiken aus Rom: Polychrome Mosaikpavimente und Emblemata aus Rom und Umgebung*. Würzburg, Germany: CompuTEAM, 1994.

Westgate 2002
Westgate, Ruth. "Hellenistic Mosaics." In *The Hellenistic World: New Perspectives*, edited by Daniel Ogden and Sylvie Le Bohec-Bouhet, 221–51. London: Classical Press of Wales and Duckworth, 2002.

Wilson 1982
Wilson, R. J. A. "Roman Mosaics in Sicily: The African Connection." *American Journal of Archaeology* 86, no. 3 (1982): 413–28.

Yegül 2000
Yegül, Fikret. "Baths and Bathing in Roman Antioch." In *Antioch: The Lost Ancient City*, edited by Christine Kondoleon, 146–51. Princeton: Princeton University Press, 2000.

Glossary

amphora
A storage vessel usually having a large oval body with a narrow neck and two or more handles extending from the body of the vessel to the rim.

apodyterium
The dressing room of a Roman bath.

bouleuterion
A council building or assembly hall.

caesti
Boxing gloves composed of strips of leather weighted with lead or iron and wrapped around the hands and forearms.

caldarium
The hot room in a Roman bath, sometimes containing a hot-water plunge.

chlamys
A short cloak or wrap worn by men.

emblema (plural *emblemata*)
The centerpiece of a mosaic; usually a small picture panel with figural representations set into a large floor mosaic.

frigidarium
The cold room in a Roman bath, sometimes containing a cold-water plunge.

guilloche
A decorative pattern created by two or more twisted bands or lines, often giving the effect of a braid.

kantharos (plural *kantharoi*)
A drinking vessel featuring a deep bowl set on a tall stem with two side handles extending from the bottom of the vessel to the rim.

kerykeion
A herald's staff or wand with two snakes entwined around it, often crowned by two wings. Also known as a caduceus.

krater
A large vessel with a wide mouth and two side handles, generally used for mixing wine and water.

missorium
A large plate.

nymphaion (plural *nymphaea*)
A monument dedicated to the nymphs. *Nymphaea* served as reservoirs, sanctuaries, and places of leisure.

odeion
A building or other roofed structure used for musical performances.

opus mixtum
A Roman construction technique that uses a combination of different types of brickwork.

opus sectile
A mosaic technique used to decorate floors and walls in which designs are created by fitting together pieces of stone or glass larger than the small cubes used in *tesserae*.

opus signinum
A mortar pavement made of concrete mixed with crushed tile or stone chips.

opus tessellatum
Mosaic technique using small cubes of stone or glass known as *tesserae* to form ornamental patterns and figural scenes.

opus vermiculatum
A technique used to decorate floors and walls in which tiny pieces of irregular stone are used to create extremely detailed picture mosaics.

pilae
Pillars composed of stacked tiles; used to raise the floor of a *caldarium* to allow heating from below.

situla
A bucket-shaped vessel, often having handles.

terrazzo
A concrete mixed with chips of marble or stone; used especially as a decorative surfacing on floors and walls.

tesserae
Small square or cubed pieces of hand-cut stone, glass, or terracotta used in the making of a mosaic.

thyrsus
A staff of fennel topped with a pinecone and sometimes twined with ivy leaves and vines.

triclinium
The dining room of a Roman house.

venatio (plural *venationes*)
A type of entertainment held in Roman arenas that involved the staged hunting and killing of wild animals.

Contributors

Alexis Belis is assistant curator of antiquities at the J. Paul Getty Museum.

Nicole Budrovich is curatorial assistant at the J. Paul Getty Museum.

Christine Kondoleon is the George D. and Margo Behrakis Curator of Greek and Roman Art at the Museum of Fine Arts, Boston.

Kenneth Lapatin is associate curator of antiquities at the J. Paul Getty Museum.

Sean Leatherbury is assistant professor of ancient art and culture at Bowling Green State University, Ohio.

Index

Note: page numbers in *italics* refer to images; terms in **bold** preceding the page numbers refer to sections of this catalogue.